NORTHEAST ASIA'S
NUCLEAR CHALLENGES

Explorations in Korean Studies

NORTHEAST ASIA'S
NUCLEAR CHALLENGES

Edited by
Su Hoon Lee

IFES
KYUNGNAM UNIVERSITY PRESS

Published by the Institute for Far Eastern Studies, Kyungnam University
Seoul, South Korea

Distributed outside of North and South Korea by
Lynne Rienner Publishers, Inc.
1800 30th Street, Boulder, CO 80301, USA
www.rienner.com

ISBN 978-89-8421-347-0

Contents

Tables

Figures

Introduction

A Myriad of Challenges—Carving a Peaceful, Stable, and Secure Northeast Asia

Su Hoon Lee

The actors in Northeast Asia face an array of nuclear challenges and corresponding shared tasks. The denuclearization of North Korea is one of them. Ensuring the safety of nuclear-power infrastructure from natural and man-made disasters is another. Preventing proliferation of nuclear weapons and programs is a third. Many other challenges exist, many of which are long term. The imperative is rising for these actors—that is, China, South Korea, Japan, Russia, and the United States—together with the international community to come together to cooperatively deal with these issues. Peace and security in the region and indeed worldwide will depend on our ability to establish solid mechanisms for cooperation and collective action to meet the array of challenges.

In light of the growing global concerns about these challenges as they pertain to the actors in Northeast Asia, in March 2012 the Institute for Far Eastern Studies of Kyungnam University (IFES) and the Friedrich Naumann Foundation for Liberty Korea Office (FNF) hosted an international conference titled "Rethinking Nuclear Issues in Northeast Asia." The purpose of the conference was to gain a closer expert assessment of the nuclear issues that the region faces in terms of nuclear safety and security. Contained within this volume are the papers that were prepared in anticipation of this event. Distinguished senior experts and scholars from China, Japan, Russia, South Korea, and the United States provided multidimensional perspectives on crucial issues. In many cases, policy recommendations were also offered.

The seven chapters in this volume are of varying length and style yet nonetheless coherently dissect and assertively address the political dimensions of the issues of nuclear safety and security in Northeast Asia, including the growing threats posed by state and nonstate actors to the global nuclear nonproliferation regime and the essential need to increase the efficacy of cooperative approaches and multilateral efforts to meet these threats, as well as address the debate on the safety and sustainability of nuclear power as an energy source.

In Chapter 1, Leon Sigal identifies the small-yet-positive headway that has been made by the US president, Barack Obama, in changing US nuclear policy, doctrine, practices, and programs—for the better, that is, despite the unyielding domestic and stiff international political challenges and constraints the Democratic president has faced. In assessing the critical proliferation threat that North Korea's nuclear programs continue to present the region and the global nonproliferation regime, Sigal asserts that the North Korean nuclear issue may hold "the best hope for success in curbing proliferation" globally. That hope, however, is predicated on Washington's willingness to sustain an approach of diplomatic give-and-take with Pyongyang, while acknowledging the imperative of a peace process to defuse the conflict on the Korean peninsula and induce North Korea to give up its nuclear and missile programs.

Chapter 2 provides a Russian perspective on the Global Partnership against the spread of weapons and materials of mass destruction. Alexander G. Savelyev analyzes the Global Partnership Program (GPP) established by the states of the G-8 in 2002, stressing that a high level of political factors have created an atmosphere of distrust, mostly between Russia and the United States. Despite the obstacles and the varying results in the different sectors of its implementation, Savelyev assesses the GPP to have been mostly successful and should continue for the benefit of all parties.

In Chapter 3, Sang-Hyun Lee reviews from a South Korean perspective the historical background of the G-8 global partnership and global nuclear nonproliferation regime. Lee identifies the partnership's major achievements as well as the need for the international community to work toward building a synergy effect between the GPP and the biennial Nuclear Security Summit (of which as of this writing two have been held, the first in Washington in April 2010 and the second in Seoul in March 2012). In Lee's view, nuclear terrorism is a real and urgent threat; past efforts to deal with it and the current state of the global nuclear-nonproliferation regime have been and are less than satisfactory. Rectifying this situation will be up to "the willingness of the nations that care about the future of humanity." To demonstrate that willingness, extension of the global partnership is essential. According to Lee, South Korea, for its part, seems poised to take a more active role in such collective nuclear-nonproliferation endeavors.

Looking more specifically at the regional and global security threat posed by North Korea's nuclear programs, in Chapter 4 Kun Young Park calls for a "paradigmatic" change in the minds of policymakers in South Korea and the United States. Park outlines a new "strategic-pragmatic-comprehensive" (SPC) approach to facilitate North Korean denuclearization. Park discusses the dangers posed by a nuclear North Korea, what he sees as the poverty of past non-SPC approaches to this end pursued by the Lee Myung-bak administration in Seoul and the Obama administration in Washington, and argues for policy alternatives for future US and ROK administrations based on the SPC perspective. In sum, Park believes that while North Korea is highly unlikely to relinquish its nuclear-weapons capability in the immediate future, in the longer term, by putting themselves in North Korea's shoes, Seoul and Washington will be better able to induce North Korea to negotiate a deal because cost-benefit analyses and structural constraints will urge all of these actors to travel that road.

In Chapter 5, former director of the Los Alamos National Laboratory and renowned scientist Siegfried S. Hecker shares his intimate knowledge of the North Korean nuclear programs and his views on the prospects of solving the nuclear crisis. While Hecker believes that, at this stage, preventing North Korea from substantially enhancing its nuclear program is the best the international community can hope to achieve in the short term, he, like Park, remains optimistic that the crisis can be resolved in the long term: "Paradoxically, compared to a more democratic country, an autocracy like North Korea may find it easier to give up its nuclear weapons if doing so is seen to help the regime survive, because it does not have to deal with domestic opposition." The challenge, as Hecker sees it, will be for Washington and Seoul to find a way to improve security in the region while simultaneously dealing with their own domestic politics to encourage

North Korea's transition.

Moving from the proliferation danger to the energy dilemma, in Chapter 6 Jian Cai describes the history and current status of the nuclear power industry in China, South Korea, and Japan—Northeast Asia's big three economies—and analyzes their future nuclear policies. Acknowledging that globally the role of nuclear infrastructure for power generation has peaked in many nations, Cai questions whether or not Asia is actually poised for a "nuclear renaissance," with the likes of China, India, Japan, and South Korea at the center of the global nuclear-power expansion, all driven by the growing energy demand, desire for energy security, and heightened environmental concerns. In the wake of the 2011 disaster at the Fukushima Daiichi nuclear power plant in Japan, there are renewed concerns over the safety and viability of nuclear energy. Overall, Cai concludes that economic, environmental, health, and proliferation risks implicit in nuclear power render nuclear technology a failing score in terms of meeting the criteria for a sustainable energy future. Hence, Cai calls for greater investment in clean and renewable energy sources. Nevertheless, Cai acknowledges that many countries in Asia still hold firm plans to construct new nuclear reactors for power generation. While countries in the region seemed to have made nuclear safety a greater priority following the Fukushima disaster, in the cases of China and South Korea, for example, the level of current investment in the nuclear-power sector and increasing search for energy security diminishes the prospects of a halt or slowdown in this sector in the near future.

Finally, in Chapter 7 Takao Takahara points out that the complacency that preceded the nuclear disaster at Fukushima suggests that "if one is prescient enough, the issue of nuclear weapons needs to be addressed urgently." Takahara argues that the "only way to avoid the nuclear hell which nobody wants is to delegitimize and eliminate nuclear weapons." For Takahara, the path to that goal should start with negotiations for a Nuclear Weapons Convention that would ban the use of such weapons, alongside the proposal of constructing a regional Nuclear-Weapon-Free Zone. In his view, such a goal and path could allow for meaningful cooperation between nations like South Korea and Japan.

As the chapters within this volume suggest, for the actors in Northeast Asia, carving a vision of a peaceful, stable, and secure region requires that these nuclear challenges be addressed comprehensively, cooperatively, and immediately. The outlook is opaque, and the tasks daunting. Nevertheless, hope remains that measured, incremental steps will be taken collectively, and positive outcomes achieved.

Chapter 1

An Assessment of the Obama Doctrine

Leon Sigal

How has President Barack Obama fared in his efforts to change nuclear policy, doctrine, practices, and programs? In this policy domain as in others, Obama has pursued demanding goals in an unyielding domestic and international political environment. At home, many Republicans in Congress and some career officials in the bureaucracy have balked at his initiatives. Abroad, proliferating states have posed a stiff challenge. Still, he has made headway.

The Prague Speech

In Prague on April 5, 2009, President Obama gave ringing rhetorical support to nuclear abolition, speaking of "America's commitment to seek the peace and security of a world without nuclear weapons." He acknowledged that "this goal will not be reached quickly—perhaps not in my lifetime."

The president went on: "As long as these weapons exist, the United States will maintain a safe, secure and effective arsenal to deter any adversary, and guarantee that defense to our allies." Nuclear conservatives in the US bureaucracy took those words as license to resist changes in existing policy and practices. That, in effect, reversed his priorities, which were nuclear proliferation and nuclear terrorism, not nuclear deterrence.

Highlighting nuclear terrorism as "the most immediate and extreme threat to global security," Obama called for "a new international effort to secure all vulnerable nuclear material around the world within four years." That effort has had modest success. Obama pledged "concrete steps" by the United States toward the goal of nuclear abolition. He has made good on his promises to "reduce the role of nuclear weapons in our national security strategy" and to "reduce our warheads and stockpiles" by negotiating arms reductions with Russia.

His other initiatives, which required congressional and international action, have not fared as well. He said he would seek ratification of the Comprehensive Test Ban Treaty, "immediately and aggressively" pursue "a new treaty that verifiably ends the production of fissile materials intended for use in state nuclear weapons," and strengthen the nuclear Nonproliferation Treaty (NPT) by providing more resources for inspections and by "creating a new framework for civil nuclear cooperation, including a nuclear fuel bank.

To strengthen the NPT, he also wanted to impose "real and immediate consequences for countries caught breaking the rules or trying to leave the treaty without cause." He seized on a timely example: "Just this morning, we were reminded again of why we need a new and more rigorous approach to address this threat. North Korea broke the rules once again by testing a rocket that could be used for long range missiles." He went on: "Rules must be binding. Violations must be punished. Words must mean something. The world must stand together to prevent the spread of these weapons. Now is the time for a strong international response."[1]

No doubt the nonproliferation regime needs to be preserved

---

[1] President Barack Obama, Prague Speech on Nuclear Weapons, April 5, 2009.

and strengthened, but criminalizing a state's violations is no way to do it. The crime-and-punishment approach has never kept countries from developing nuclear weapons. It has failed where diplomatic give-and-take has succeeded. As if to underscore that history, no sooner had the UN Security Council imposed sanctions on North Korea for its rocket launch than that country began preparations for a nuclear test—a test it carried out in May 2009. Without a determined effort by the United States to negotiate with North Korea and Iran, Obama's efforts to curb proliferation will fall short. As National Security Adviser Thomas Donilon has characterized the situation, "Unless we can meet the challenge posed by Iran and North Korea, additional countries in the Middle East and East Asia could well leave the NPT and develop their own nuclear weapons, thus reversing any movement towards disarmament."[2]

With these general observations in mind, it might be useful to move beyond President Obama's statements of nuclear policy and examine his nuclear doctrine, practices, and programs in greater detail, with special attention to nonproliferation in Northeast Asia.

Beyond Nuclear Policy to Nuclear Doctrine

Lofty as the aims of his Prague speech were, Obama has taken some practical steps to realize them—most notably, intervening to alter a draft Nuclear Posture Review that nuclear conservatives in the bureaucracy attempted to foist on him. They had long assumed that whatever the United States did with its own nuclear weapons had no bearing on proliferation. Obama's Nuclear Posture Review has a very different

---

2 Thomas Donilon, Keynote Address to the 2011 Carnegie International Nuclear Policy Conference, March 29, 2011.

premise: that reduced numbers of US nuclear weapons and reduced reliance on them can help mobilize global support for US nonproliferation goals, even if it does not get Iran and North Korea to change course.

The Obama Nuclear Posture Review stipulates a receding role for nuclear arms in US defense strategy:

> The United States will continue to strengthen conventional capabilities and *reduce the role of nuclear weapons in deterring non-nuclear attacks*, with the objective of making deterrence of nuclear attack on the United States or our allies and partners the sole purpose of US nuclear weapons.[3]

Fulfilling the promise of Prague will depend on "follow-on analysis to set goals for future nuclear reductions below the levels expected in New START," which is to be completed in coming days.

The Nuclear Posture Review seeks to reassure China and Russia about US nuclear intentions. It commits Washington to "pursue high-level bilateral dialogue" with Beijing and Moscow aimed at fostering "a more stable, resilient, and transparent strategic relationship."[4] By repeatedly affirming the goal of strategic stability, it acknowledges China's nuclear deterrent and eschews a disarming first-strike against its small arsenal.

The centerpiece of the 2010 Nuclear Posture Review is a firmer negative security assurance, made possible by the reduced US reliance on nuclear weapons:

> With the advent of US conventional military preeminence and continued improvements in US missile defenses and capabilities to counter and

---

[3] US Department of Defense, *Nuclear Posture Review Report*, April 6, 2010, p. ix. Emphasis added.

[4] Ibid., p. x. Emphasis added.

> mitigate the effects of CBW [chemical and biological weapons], the role of US nuclear weapons in deterring nonnuclear attacks—conventional, biological, or chemical—has declined significantly. The United States will continue to reduce the role of nuclear weapons in deterring nonnuclear attacks. To that end, the United States is now prepared to strengthen its long-standing "negative security assurance" by declaring that *the United States will not use or threaten to use nuclear weapons against non-nuclear weapons states that are party to the NPT and in compliance with their nuclear non-proliferation obligations.*[5]

This assurance usefully supplants the one announced by Secretary of State Cyrus Vance at the UN Special Session on Disarmament on June 12, 1978:

> The United States will not use nuclear weapons against any non-nuclear weapons state party to the N.P.T. ... except in the case of an attack on the United States, its territories or armed forces, or its allies, by such a state allied to a nuclear weapons state, or associated with a nuclear weapons state in carrying out or sustaining the attack.

The Vance exception applied to members of the Warsaw Pact and North Korea, which then was allied to the Soviet Union and China.

According to the 2010 Nuclear Posture Review, no longer will the United States plan for nuclear retaliation for other than a nuclear attack on itself or its allies:

> In making this strengthened assurance, the United States affirms that any state eligible for the assurance that uses chemical or biological weapons against the United States or its allies and partners would face the prospect of a devastating conventional military response—and that any individuals responsible for the attack, whether national leaders or military commanders, would be held fully accountable.[6]

---

[5] US Department of Defense, *Nuclear Posture Review Report*, April 6, 2010, p. viii.

[6] Ibid. Emphasis added.

> While the Nuclear Posture Review marks a significant advance in US declaratory policy, it falls short of the declaration of no first use of nuclear weapons or an alternative formulation proposed by Morton Halperin and Arnold Kanter: "The United States maintains nuclear weapons to deter, and if necessary, respond to *nuclear attacks* against itself, its forces, or its friends and allies."[7]

Yet the Obama language stands in sharp contrast to the expansive view of nuclear requirements contemplated in the 2002 Nuclear Posture Review completed under President George W. Bush:

> In setting requirements for nuclear strike capabilities, distinctions can be made among the contingencies for which the United States must be prepared. Contingencies can be categorized as immediate, potential or unexpected. Immediate contingencies involve well-recognized current dangers.... Current examples of immediate contingencies include an Iraqi attack on Israel or its neighbors, a North Korean attack on South Korea, or a military confrontation over the status of Taiwan.[8]

The implications of the shift from Bush to Obama for North Korea were explicit. In the words of the Obama Nuclear Posture Review,

> This revised assurance is intended to underscore the security benefits of adhering to and fully complying with the NPT and persuade non-nuclear weapon states party to the Treaty to work with the United States and other interested parties to adopt effective measures to strengthen the non-proliferation regime.... *In the case of countries not covered by this assurance—states that possess nuclear weapons and states not in compliance with their nuclear non-proliferation obligations—there remains a narrow range of contingencies in which US—nuclear weapons may still play a role in deterring a conventional or CBW attack against the United States or its allies and partners.*[9]

---

[7] Morton H. Halperin, "Promises and Priorities," *Survival* 51, no. 5 (October–November 2009): 20.

[8] US Department of Defense, *Nuclear Posture Review*, December 31, 2002, p. 16.

[9] US Department of Defense, *Nuclear Posture Review Report*, April 6, 2010, p. viii. Emphasis added.

The language strongly implies that even in the case of North Korea, while the United States was not forswearing the use of nuclear weapons, neither was it threatening to use them.

North Korea was not reassured. Its response to the 2010 Nuclear Posture Review was to issue a nuclear policy declaration of its own:

> President Obama blustered that the US will not use or threaten to use nuclear weapons against non-nuclear weapons states that comply with the provisions of NPT but exception is made for countries such as the DPRK and Iran. This proves that the present US policy towards the DPRK is nothing different from the hostile policy pursued by the Bush administration at the outset of its office during which it was hell-bent on posing a nuclear threat to the DPRK after designating it as a "target of preemptive nuclear strike." By releasing the review the US completely backpedaled its commitment made in the September 19 Joint Statement of the six-party talks that it has no intention to attack or invade the DPRK with nuclear weapons or conventional weapons, and again chilled the hard-won atmosphere for the resumption of the talks.... The DPRK has so far sincerely implemented its international obligation as a responsible nuclear weapons state. The denuclearization of the Korean Peninsula is the invariable goal of the DPRK. If the peninsula and the rest of the world are to be denuclearized, the US should stop such hostile acts as trampling down upon other countries' sovereignty and right to existence, pursuant to its policy of strength based on nuclear supremacy. *What is most urgent is for the US to roll back its hostile policy towards the DPRK in practice, not with an empty talk, and take a confidence-building measure. As long as the US nuclear threat* persists, the DPRK will increase and update various type nuclear weapons as its deterrent in such a manner as it deems necessary in the days ahead. The DPRK is fully capable of doing so. It is the US that gives the former ground and justification to do so.[10]

---

[10] KCNA, "Foreign Ministry Dismisses US Nuclear Plan," April 9, 2010. Emphasis added.

What stands out is North Korea's repeated emphasis on a fundamentally new relationship with the United States. The implication is that reconciliation—an end to enmity or what the North calls US "hostile policy"—would mean an end to the nuclear threat even if the United States still retained nuclear weapons.

Beyond Nuclear Doctrine to Nuclear Practice

Nuclear doctrine sets the parameters for the armed services and the national laboratories to bargain over nuclear forces, programs, and budgets. In these areas, the Obama legacy so far has been mixed.

Deep Cuts

The Obama administration has negotiated and the Senate has approved New START with Russia, lowering the ceiling on deployed strategic nuclear warheads 30 percent to 1,500. Although Moscow was prepared to accept a somewhat lower ceiling than Washington, much deeper reductions will require more binding assurances to Moscow about the scope and technical character of missile defense deployments. Deferring to the wishes of Turkey and Poland, however, Washington did not withdraw its remaining nonstrategic or tactical nuclear weapons from Europe—some 200 B-61 gravity bombs. Instead it tied their removal to negotiations with Moscow on a common ceiling for all warheads, strategic and nonstrategic, deployed and nondeployed. Whether Congress will fund a program to extend the life of these old bombs or the Europeans will replace their aging F-15 and F-16 fighter-bombers to deliver them is uncertain.

Comprehensive Test Ban Treaty (CTBT)

In seeking Senate ratification of the Comprehensive Test Ban Treaty, the Obama administration made a Faustian bargain with Republicans led by Senator John Kyl: a substantial increase in funding for weapons work at the national laboratories, including refurbishing aging warheads to assure their safety and reliability. At the same time, however, the Republican-dominated House of Representatives cut back funding for the labs' essential work on nonproliferation. The net result so far is no CTBT and a nuclear budget out of balance, with expenditures to maintain and improve nuclear weapons dwarfing those on dismantling and disposing of old warheads, detecting and analyzing proliferation, and verifying treaties.

Fissile Material Cut-Off Treaty (FMCT)

A treaty to cut off global production of fissile material has long been bogged down in the Conference on Disarmament (CD) and Pakistan and India are generating much more. More positively, the CD broke a twelve-year deadlock of its work program last May, agreeing by consensus to negotiate a treaty, and the United States, Russia, France, and the UK have all ceased production unilaterally.

Nuclear Safety and Security

To date, administration efforts have focused on the risk that terrorists could somehow seize a nuclear weapon or enough fissile material to make one, a risk that combines low probability with appallingly high cost. In keeping with that focus, putting an end to civil use of highly enriched uranium around the world is the subject of the Nuclear Security Summit to be held in Seoul on March 26–27, 2012.

Yet far less attention has been devoted to a much more likely threat—that terrorists could make a "dirty bomb," using conventional explosives and radioactive material that can be found in hospitals and university labs in Seoul and most other cities around the globe. Detonating such a device could shut down a tunnel or bridge for many months and create havoc in populated areas. The issue of radiological security was belatedly but appropriately added to the Seoul summit agenda.

Curbing Proliferation

The main threat to President Obama's goal of abolition comes from a small group of states that are poised to make more nuclear weapons. Iran has received the most attention recently, but Pakistan may pose a greater threat to the nonproliferation regime. It has exported nuclear technology and know-how to North Korea and Libya, among other countries, and may yet try to do so to Saudi Arabia in the future. With new fissile material production facilities now in operation, this unstable state is on a trajectory to have the world's fourth largest stock of nuclear weapons. That surge has not passed unnoticed in India. A nuclear arms race in South Asia could compel China to expand its arsenal, which is much smaller than is commonly acknowledged. This, in turn, could generate pressures for nuclear proliferation in Northeast Asia.

By contrast, Iran's nuclear weapons ambitions are far from clear. In its latest report to Congress on global threats, the US intelligence community says:

> We assess Iran is keeping open the option to develop nuclear weapons, in part by developing various nuclear capabilities that better position it to produce such weapons, should it choose to do so. We do not

> know, however, if Iran will eventually decide to build nuclear weapons.[11]

The latest report of the International Atomic Energy Agency confirmed earlier US intelligence assessments of "possible military dimensions," saying that such activities had largely ceased. The IAEA also disclosed that, as of February 11, 2012, Iran had enriched 95.4 kilograms of uranium hexafluoride to 19.75 percent.[12] IAEA data show Iran has been increasing its stock of low-enriched uranium, but at a reduced rate, suggesting that it is having difficulty operating its first-generation centrifuges and has yet to perfect its second-generation ones. At its current pace, it would take two years or more for Tehran to enrich enough uranium to weapons grade for a bomb if it chose to do so—and still longer to fabricate a weapon—leaving ample time for negotiations.

In his 2012 State of the Union address, President Obama emphasized coercive measures, while holding out the opportunity for negotiation:

> America is determined to prevent Iran from getting a nuclear weapon, and I will take no options off the table to achieve that goal. But a peaceful resolution of this issue is still possible, and far better, and if Iran changes course and meets its obligations, it can rejoin the community of nations.[13]

Obama's statement, however, seemed to rule out one option: deterring Iran if it were to arm itself with nuclear weapons.

---

[11] Director of National Intelligence James R. Clapper, *Unclassified Statement for the Record on the Worldwide Threat Assessment of the US Intelligence Community*, US Senate, Select Committee on Intelligence hearing, January 31, 2012.

[12] IAEA Director General, Report to the Board of Governors, *Implementation of the NPT Safeguards Agreement and Relevant Provisions of the Security Council Resolutions in the Islamic Republic of Iran*, February 24, 2012, p. 5.

[13] President Barack Obama, *State of the Union Address*, January 24, 2012.

Whether a negotiated resolution is possible given the current power struggle in Tehran is an open question. The US intelligence community's assessment leaves open the possibility: "We judge Iran's nuclear decision-making is guided by a cost-benefit approach, which offers the international community opportunities to influence Tehran."[14]

The Bush administration had rejected an Iranian opening to comprehensive talks in 2003 and repeatedly stymied efforts by the European Union to negotiate with Iran. By contrast, in 2009 the Obama administration, facing down opposition from Israel, Saudi Arabia, and Congress, tried and failed to get Iran to swap low-enriched uranium it had produced in return for uranium fuel for its isotope-generating research reactor. Washington, however, blocked a subsequent Turkish-Brazilian attempt to negotiate with Tehran.[15] At this point, allowing Iran to enrich uranium under close international monitoring to ensure it stops short of producing the highly enriched uranium needed for weapons may be the only viable negotiated outcome.

Amid the resounding drumbeat for military action, much of it from the same quarters that pressed for the disastrous second war with Iraq, the administration has focused on mobilizing global support for tougher economic sanctions. Sanctions may prove ineffective. They are likely to enrich the Republican Guards, who control the country's smuggling, and empower hardliners who oppose yielding to pressure and can appeal to Iranian nationalism.

Coercion alone has never stopped proliferators. Negotiation, by contrast, has had success in South Korea, Taiwan, Sweden, Brazil, Argentina, South Africa, Ukraine, Belarus, Kazakhstan, and, most recently, Libya.

---

[14] Clapper, *Statement for the Record.*

[15] Trita Parsi, *A Single Roll of the Dice* (New Haven: Yale University Press, 2012), chs. 8, 10. Cf., Mohamed ElBaradei, *The Age of Deception* (New York: Henry Holt, 2011), ch. 11.

Nuclear Diplomacy with North Korea

North Korea may yet hold out the best hope for success in curbing proliferation, but only if President Obama sustains diplomatic give-and-take. It took nearly three years for his administration to negotiate in earnest with North Korea. In the meantime, North Korea conducted a second nuclear test, showed off one new, longer-range missile and tested another, and revealed a surprisingly advanced 2,000-centrifuge enrichment facility at Yongbyon to visiting Americans.

It has yet to enrich significant quantities of weapons-grade uranium or restart its five-megawatt reactor at Yongbyon, complete construction of a new light-water reactor, and conduct the additional nuclear and missile tests it needs to develop a new miniaturized warhead and more reliable missiles to deliver it.

Continued arming by North Korea could have grave consequences for security in Northeast Asia. While such a development would shore up US alliances in the near term, it may eventually sow doubts in South Korea and Japan about relying on the United States for their security. Already some in Seoul have called for the return of US nuclear weapons to the peninsula—or worse, resuming a South Korean nuclear program that Washington stopped twice before. It will also strengthen the hand of right-wing nationalists in Japan who distrust the United States and favor nuclear-arming. Worst of all, it will generate increased political pressure to confront China for failing to bring North Korea to its knees. That could jeopardize cooperation between Washington and Beijing, the key to allied security.

Given the risks to Northeast Asia, as well as to President Obama's Prague goals, stopping North Korea's nuclear programs is a matter of urgency. It is one thing to live with a North Korea that has a few nuclear devices best delivered by container ship and quite another to have an unbounded nuclear arsenal, deliverable by missiles.

Diplomatic give-and-take may yet bear fruit. Just before he died, Kim Jong-il was on the verge of concluding a deal with the United States to suspend enrichment at Yongbyon with IAEA monitoring in return for food aid, which Pyongyang called a "confidence-building measure"—a reassuring sign that Washington was easing hostility. He also seemed willing to declare a moratorium on nuclear and missile tests while negotiations proceeded. The North has also indicated readiness to resume talks with South Korea to ship out the fuel rods needed to restart its Yongbyon reactor in return for energy aid. But permanent dismantlement of its nuclear and missile programs, never mind its weapons, will require much more substantial political and economic engagement.

Some question the utility of negotiations. They claim that the DPRK will exploit talks to seek recognition as a nuclear-weapons state. Pyongyang has not asked for such recognition, and Washington will never grant it. Skeptics say North Korea has always been determined to nuclear-arm and that talks were a ploy to buy time for arming. Yet a review of the negotiating history casts doubt on that view. It also suggests how counterproductive sanctions and other coercive measures have been.

A Checkered Past

Until recently, the only way for the North to make the fissile material it needed for weapons was to remove spent nuclear fuel from its reactor at Yongbyon and reprocess it to extract plutonium. Yet North Korea stopped reprocessing in the fall of 1991, some three years before signing the 1994 Agreed Framework, and did not resume until 2003. Whether it had enough fissile material for a bomb at the time is still in doubt, but the halt kept it from making many bombs' worth more.

The North did begin acquiring the means to enrich uranium in

1997—but only after the Clinton administration was slow to live up to its end of the Agreed Framework. It did not restart its plutonium program, however. It took that fateful step in 2003 only after the Bush administration confronted it in October 2002 over its acquisition of enrichment equipment and refused its offer to negotiate. It also stepped up enrichment acquisitions and proclaimed itself a nuclear-armed state.

Today, again, the five-megawatt reactor at Yongbyon remains shut down as agreed in six-party talks in February 2007. The North could restart it in a manner of months. The reactor could produce a bomb's worth of fissile material sooner than the nearby enrichment plant.

Similarly, the conventional wisdom in Washington is that President Obama held out his hand to Kim Jong-il only to have it slapped away. Again, history suggests otherwise. In the interregnum between the 2008 election and his inauguration, consumed by the financial meltdown and looming depression, President-elect Obama decided to strengthen ties with South Korea and Japan instead of resuming talks with North Korea. Administration officials were aware that the South's president, Lee Myung-bak, and the then—prime minister, Aso Taro, were determined to isolate and pressure Pyongyang in hopes of making it more pliable in negotiations, if not compelling its collapse. The US embassy in Seoul warned Washington of what it called Lee's "tougher approach" in a January 29, 2009 cable disclosed by WikiLeaks:

> President Lee ... is prepared to leave inter-Korean relations frozen until the end of his term in office, if necessary. It is also our assessment that Lee's more conservative advisors and supporters see the current standoff as *a genuine opportunity to push and further weaken the North, even if this might involve considerable brinkmanship*.[16]

---

[16] SEOUL 000059, Subject: ROK'S Foreign Policy toward the Neighbors: North Korea, Japan, China, and Russia, January 12, 2009. Emphasis added.

Wary of Obama, hardliners in Seoul and Tokyo were making common cause. If Washington and Pyongyang were to speed up bilateral talks, one senior South Korean official told *Chosun Ilbo*, Japan could play a role in "slamming on the brakes."[17]

South Korea and Japan had already slammed on the brakes by undoing the October 2007 six-party, second-phase agreement under which Pyongyang had pledged to provide "a complete and correct declaration of all its nuclear programs" and to disable its plutonium facilities, pending their permanent dismantlement. In return, it was promised energy aid, an end to US sanctions under the Trading with the Enemy Act, and removal from the US list of state sponsors of terrorism. The accord made no mention of verification, which was left to a later phase of negotiations. Under pressure from Seoul and Tokyo, President Bush delayed the delisting and sanctions easing until Pyongyang agreed to cooperate in verifying its nuclear declaration. As Secretary of State Condoleezza Rice acknowledged on June 18, 2008, Washington had moved the goalposts: "What we've done, in a sense, is move up issues that were to be taken up in phase three, like verification, like access to the reactor, into phase two."[18] Even after Pyongyang accepted arrangements that could have sufficed to ascertain how much plutonium it had extracted prior to 1991,[19] Seoul suspended delivery of promised fuel oil, and Bush backed it.

Upon taking power, the Obama administration sustained the Bush course. It did not undo the renege on fuel aid. Its posture of "strategic patience" left its North Korea policy hostage to its allies,

---

[17] *Chosun Ilbo*, "Japan 'Could Become Seoul Ally in N. Korea Issues,'" March 24, 2009.

[18] Secretary of State Condoleezza Rice, "US Policy towards Asia," Address at the Heritage Foundation, June 18, 2008.

[19] US Department of State, Office of the Spokesman, US–North Korea Understandings on Verification, October 11, 2008.

which were doing their utmost to impede nuclear diplomacy. Indeed, Obama would go longer without direct talks with North Korea than Bush had. Disengagement by Washington and Seoul has never gone down well in Pyongyang. The most dangerous crises have come when Seoul has blocked engagement between Washington and Pyongyang. Even worse, with transition to a new generation of leadership under way in Pyongyang, loose talk about impending collapse of the North Korean regime was ominously rife in Seoul.

In late January 2009, Pyongyang moved to force the action by assembling a rocket at its launch site at Musudan-ri. It took two months, giving Washington time to resume talks. It tried to portray the test launch as a peaceful attempt to put a satellite into orbit, but in private it made clear to visitors that without the promised energy aid it would have no recourse but to strengthen its deterrent. On April 5, 2009, it launched the rocket. Within days, the UN Security Council, in a presidential statement drafted by the United States and China, condemned the launch and imposed sanctions.

Spurning the UN action, a DPRK Foreign Ministry spokesman denounced six-party talks as "an arena which infringes on our sovereignty and which aims only at disarming us and overthrowing our system" and said Pyongyang "will no longer be bound by any agreement." That called into question its 2005 commitment to "abandon" its nuclear weapons and existing programs. The spokesman listed four other steps Pyongyang would take in response. First, "we will actively examine the construction of a light-water [nuclear] plant of our own."[20] Such a plant would require enriched uranium to fuel it. Second, "we will strengthen our self-defensive nuclear deterrent in every way," which some took to mean it would conduct another

---

20 KCNA, "DPRK Foreign Ministry Vehemently Refutes UNSC's 'Presidential Statement'," April 14, 2009.

nuclear test. Third, the Yongbyon facilities "will be restored to the original state for normal operation," which stopped short of saying North Korea would restart its reactor to generate more spent nuclear fuel. And fourth, the 6,500 spent fuel rods removed during disabling "will be reprocessed."

No sooner did it issue the statement than Pyongyang began preparations for its second nuclear test—a test it conducted on May 25. That prompted additional UN and US sanctions. North Korea's response was to reveal its enrichment facility at Yongbyon, underscoring the futility of sanctions in curbing its nuclear programs. With negotiations slow to begin, US Secretary of Defense Robert Gates exhibited strategic impatience. On January 11, 2011, while on a trip to Asia, he told reporters that the status quo in Korea had changed in two significant ways: "The first is, with the North Koreans' continuing development of nuclear weapons, and their development of intercontinental ballistic missiles, North Korea is becoming a direct threat to the United States."—The second "is clearly … if there is another provocation, there will be pressure on the South Korean government to react"—a diplomatic way of indicating US military unease with Seoul's aggressive stance. He concluded,—"We think there is *some urgency to proceeding down the track of negotiations and engagement*."[21]

Yet it would be summer 2011 before Washington moved. In three rounds of bilateral talks with Washington, Pyongyang confirmed that it was prepared to suspend uranium enrichment and keep its reactor shut down with international monitoring and that it was willing to abide by a moratorium on nuclear and missile tests while "productive dialogues continue." It also reaffirmed that "the 1953 Armistice Agreement is the cornerstone of peace and stability on the

---

[21] US Department of Defense, Secretary of Defense Robert M. Gates, Media Roundtable from Beijing, January 11, 2011. Emphasis added.

Korean Peninsula until the conclusion of a peace treaty."[22] In return, the United States reaffirmed the October 2000 joint statement that "it does not have hostile intent toward the DPRK and is prepared to take steps to improve our bilateral relations," agreed to provide "an initial 240,000 metric tons of nutritional assistance with the prospect of additional assistance based on continued need," and pledged to increase cultural, educational, and sports exchanges.[23] The commitments by the two sides were made public on February 29, 2012.

Left unresolved was whether the missile test moratorium precluded satellite launches. That is important because the first two stages of the rocket North Korea uses to put a satellite into orbit are indistinguishable from a longer-range missile to deliver a nuclear warhead. The North's negotiators insisted it has a sovereign right to launch satellites despite a UN Security Council ban. US negotiators responded that a satellite launch would be a deal breaker.

North Korea has long said that, given the deep mistrust between it and the United States, action for action is essential for confidence building. In the immediate aftermath of the February 29 accord opening the way to an end to hostile relations, a rocket launch would be confidence destroying. Unless it is suspended, fruitful dialogue will come to an end. That would be a shame because the arrangements of February 29 could buy at least a year's respite—time for six-party talks to make further headway. The North indicated that once those talks resumed, it would initially seek sanctions easing and the light-water reactor it was promised in the Agreed Framework but never received. Yet permanent elimination of the North's nuclear programs will require nothing short of steps to end enmity, or what Pyongyang

---

22 KCNA, "DPRK Foreign Ministry Spokesman on the Result of DPRK-US Talks, February 29, 2012.

23 Department of State, Press Statement, US-DPRK Bilateral Discussions," February 29, 2012.

calls the US "hostile policy," as well as reconciliation with South Korea and Japan.

As if to underscore the point, a DPRK Foreign Ministry report issued on February 15, 2012, the 70th anniversary of Kim Jong-il's birth, cites five agreements as the legacy he passed on to his successor: from year 2000, the June 15 North-South Declaration and the DPRK-US Joint Communiqué of October 12; the DPRK Japan Pyongyang Declaration of September 17, 2002; the September 19, 2005 Six-Party Joint Statement; and the Declaration for the Development of the North-South Relations and Peace and Prosperity of October 4, 2007. It made no explicit mention of its nuclear weapons.

The Key to Denuclearization: A Peace Process in Korea

If talks resume, foremost among the steps needed is a peace process on the Korean peninsula. For more than two decades, Pyongyang has said it wants a peace treaty to replace the armistice that terminated the Korean War. Such a treaty would be a manifestation of reconciliation with Washington and Seoul. It is inconceivable that Pyongyang would give up its nuclear and missile programs, never mind its nuclear weapons, without peace on the peninsula. As long as the United States and South Korea remain its foes, it will feel threatened and want deterrents to counter that threat. A peace process is thus critical to denuclearization.

A peace process is also essential to defuse the armed confrontation in Korea that has erupted in three firefights in recent years.

Avoiding Another Cheonan

On August 10, 2011, a brief firefight broke out between North and South Korea in the West (Yellow) Sea. The initial South Korean news agency report was carefully hedged:

> South Korea fired three shots towards the tense western sea border after one North Korean shell apparently fell near there, the military here said. The Joint Chiefs of Staff (JCS) said the South's Navy heard North Korea fire three artillery shots toward the Northern Limit Line (NLL) around 1 p.m. and then responded around 2 p.m. with three warning shots. "We estimated that one North Korean shell dropped near the NLL," a JCS official said. "We haven't noticed any particular movements in the North Korean military but we're maintaining a defense posture."[24]

It is not clear if any of the original rounds fell south of the NLL or where South Korea's return shots fell. A few more rounds were exchanged later in the day. Washington and Seoul were quick to accuse North Korea of a provocation. An annual joint US–South Korean exercise to begin this week was scarcely mentioned.

The waters off Korea's west coast have been troubled ever since the end of the Korean War in 1953, when the US Navy unilaterally imposed a ceasefire line at sea north of the Military Demarcation Line (MDL) on land. North Korea has long objected to this so-called Northern Limit Line, which is not recognized internationally. It wants the MDL line extended out to sea.

To speak of North Korean provocations is to ignore the precarious military balance in Korea and the parlous political context on the peninsula at this time. North Korea's armed forces are inferior to

---

[24] Yonhap, "S. Korea Returns Fire after N. Korean Shell Falls near Sea Border," August 10, 2011.

South Korea's—on land, in the air, and at sea. Although a North Korean surprise attack cannot be ruled out, the allies are confident they would defeat it decisively. Any use of nuclear arms by the North would be nothing short of suicidal. Deterring smaller-scale encounters is much more difficult. For its part, Pyongyang can credibly threaten devastation of much of Seoul within range of its artillery and short-range missiles, which should suffice to deter an attack by South Korea. In short, while mutual deterrence makes the likelihood of deliberate aggression on the peninsula quite low, the very steps that each side has taken to deter premeditated war have increased the risk of deadly clashes, if not inadvertent war.

Politically, current tensions date back to 2008, after South Korean President Lee Myung-bak backed away from a potentially far-reaching summit agreement signed by his predecessor with North Korea's leader Kim Jong-il in October 2007 that included a pledge "to discuss ways of designating a joint fishing area in the West Sea to avoid accidental clashes and turning it into a peace area and also to discuss measures to build military confidence."[25] Determined to get tough with North Korea, Lee abandoned engagement and prevailed on Washington to do likewise.

Lee's renege drew a bristling response from Pyongyang. In late March 2008, after building up its shore artillery near the West Sea, it accused South Korean naval vessels of violating "its" territory and launched short-range missiles into the disputed waters, underscoring the risks of leaving the issue unresolved. It also called for a permanent peace treaty to replace the armistice agreement, a step that the Lee government was loath to take.

A heated war of words erupted in 2009. On January 17, assailing

---

[25] Declaration on the Advancement of South-North Relations, Peace, and Prosperity, October 4, 2007.

South Korea's defense minister "for making full preparations for the possible third West Sea skirmish," a North Korean military spokesman warned: "We will preserve … the extension of [the] MDL in the West Sea already proclaimed to the world as long as there are ceaseless intrusions into the territorial waters of our side in the West Sea."[26] Not to be outmuscled, South Korea's defense minister told the National Assembly a month later that it "will clearly respond to any preemptive artillery or missile attack by North Korea" in the contested waters.[27] To the armed forces on both sides, the message was clear: shoot first and ask questions later.

In August 2009, Pyongyang reached out to reengage with Seoul. Kim Jong-Il sent his two top aides dealing with North-South relations to Seoul for Kim Dae-jung's funeral with a personal invitation for Lee to a third North-South summit meeting. But Lee, mistaking the gesture as a sign of weakness, spurned the invitation.

With little to show for his efforts to reengage, Kim Jong-il turned up the heat. On October 15, the North Korean Navy accused the South of sending sixteen warships into the contested waters. A report by North Korea's Korean Central News Agency said: "The reckless military provocations by warships of the South Korean Navy have created such a serious situation that a naval clash may break out between the two sides in these waters."[28] On November 9, 2009, just such a clash took place—precisely what the 2007 summit accord had sought to forestall. After a North Korean patrol boat crossed the NLL, a South Korean naval vessel fired warning shots at it. The North

---

[26] KCNA, "Principled Stand of KPA to Defend Socialist Country as Firm as Iron Wall Clarified," January 17, 2009.

[27] Yonhap, "S. Korea to Strike Back if N. Korea Provokes Armed Clash: Defense Minister," February 20, 2009.

[28] KCNA, "Halt to Intrusion of S. Korean Warships in to DPRK Waters Demanded," October 15, 2009.

Koreans returned fire and the South Koreans opened up, severely damaging the North Korean ship and causing an unknown number of casualties, including at least one death. What prompted the North Korea patrol boat to respond to the warning shots? Did the crew believe it was under attack?

North Korea's attack on the *Cheonan* was reprisal for the November 9 incident. On November 12, after Pyongyang's demand for an apology went unanswered, the party newspaper, *Rodong Sinmun*, warned it would avenge the attack: "The South Korean forces will be forced to pay dearly for the grave armed provocation perpetrated by them in the waters of the north side in the West Sea of Korea."[29] Five days later, according to North Korean accounts, Kim Jong-il went to a naval base with his high command and ordered the training of a "do-or-die unit of sea heroes."[30] That order was carried out on March 26 with the attack on the *Cheonan*, killing forty-six South Korean seamen, an attack for which Pyongyang has since denied responsibility. Why did Pyongyang opt for reprisal? The most obvious reason is that given the precarious state of the military balance from its vantage point, it needed to restore deterrence.

More trouble was in the offing. Seoul moved to bolster deterrence by conducting a series of military exercises. Trying to reassure Seoul yet restrain it, the United States joined in some of the exercises — but not those in the contested waters of the West Sea. On August 9, 2010, the North Korean artillery fired 110 rounds into those waters after the South completed a five-day exercise in the area. On October 29, a North Korean soldier at a guard post at the DMZ dropped his

---

[29] KCNA, "S. Korea Will Be Forced to Pay Dearly for Armed Provocations," November 12, 2009.

[30] Naval officer Kim Kwang-il, interviewed on Korean Central Television, April 25, 2009, quoted in Lee Young-jong and Ser Myo-ja, "Fleet Officer Says Kim Intensified Navy Training after Nov. 10 Defeat," *JoongAng Ilbo*, May 6, 2010.

machine gun, which misfired, and South Korean troops returned fire. Later that fall, after Seoul announced it would conduct a series of live-fire exercises in the West Sea, Pyongyang used the hotline to warn it not to proceed. When the South Koreans, undeterred, went ahead on November 23, the North Koreans, instead of striking in the heat of the moment, waited for more than an hour before shelling Yeonpyeong Island. The delay suggests that Pyongyang's response was deliberate, countering South Korean muscle-flexing with more of its own to bolster deterrence. The ensuing artillery exchange killed four South Koreans, two of them civilians, and an unknown number of North Koreans.

Events 2011 revealed just how precarious the military situation has become. After misidentifying an Asiana Airlines jet as one of the North's jet fighters, South Korean marines fired at the airliner, carrying 119 passengers and crew, on its flight from China as it made its descent to Incheon Airport on June 17, 2011. The plane landed unharmed. The brief firefight in August 2011 was the latest pernicious interaction.

Washington and Seoul have been full of talk about the need to deter future *Cheonans*, but it is not clear how they can do so. To the contrary, steps taken to bolster deterrence can—and do—provoke clashes. In short, deterrence alone will not keep the peace on the peninsula. The only way out is a peace process.

Step by Step Toward Peace in Korea

Negotiating a peace treaty is a formidable task. To be politically meaningful, it would require rectification of permanent land and sea borders and normalization of diplomatic and economic relations. To be militarily meaningful, it would require changes in force postures and war plans that pose excessive risks of unintended war on both sides of the demilitarized zone. That would entail redeployment of the North's forward-deployed artillery and short-range missiles to the rear, out of range of Seoul.

A peace treaty is unlikely in the absence of a more conducive political environment. One way to improve that environment is a peace process, concluding a series of interim peace agreements as stepping stones to a treaty. Such agreements, with both the United States and South Korea as signatories, would constitute token recognition of DPRK sovereignty. Provisions of the inter-Korean summit declarations of June 2000 and October 2007 can serve as a starting point.

These accords could be linked to steps toward denuclearization. Such linkage was envisioned by the six parties in the September 2005 Joint Statement, which read: "The directly related parties will negotiate a permanent peace regime on the Korean Peninsula at an appropriate separate forum." A first step could be what Seoul once called a "peace declaration." Signed by the United States, and North and South Korea, and perhaps China, Japan, and Russia as well, such a declaration could declare an end to enmity by reiterating the language of the October 12, 2000 US-DPRK joint communiqué, which stated that "neither government would have hostile intent toward the other" and confirmed "the commitment of both governments to make every effort in the future to build a new relationship free from past enmity." It could also commit the three parties, joined by China, to commence a peace process culminating in the signing of a peace treaty. The declaration could be issued at a meeting of six-party foreign ministers.

A second step long sought by Pyongyang is a "peace mechanism" to replace the Military Armistice Commission set up to monitor the ceasefire at the end of the Korean War. This peace mechanism could serve as a venue for resolving disputes like the 1994 North Korea downing of a US reconnaissance helicopter that strayed across the DMZ or the 1996 incursion by a North Korean spy submarine that ran aground while dropping off agents. The peace mechanism would include the United States and the two Koreas—the three parties that have armed forces on the ground in Korea. China, which would be a signatory to any peace treaty, might opt to participate as well. The peace mechanism could also be a forum for negotiating a series of confidence-building measures. The joint fishing area in the West Sea is one. Naval confidence-building measures such as "rules of the road" and a navy-to-navy hotline are also worth pursuing.

Lacking satellite reconnaissance, North Korea has conducted surveillance by infiltrating agents into the South. An "open skies" agreement allowing reconnaissance flights across the DMZ by both sides might reduce that risk. A more-far-reaching arrangement might be to set up a joint North-South watch center that would download real-time data from US or Japanese reconnaissance satellites. In return, North Korea would have to reciprocate step by step by permanently dismantling its nuclear and missile programs.

Other Inducements for Denuclearization

In addition to a peace process, there are other potential inducements for further denuclearization:

Energy Aid

The North's stated rationale for uranium enrichment is to make fuel for the new light-water reactor it is building at Yongbyon to replace the two light-water reactors needed to generate electricity that were promised in the 1994 Agreed Framework but never delivered. The North has renewed its demand for those reactors in return for moving beyond suspension of enrichment to permanent dismantlement. The September 2005 six-party joint statement addresses the issue: "The

DPRK stated that it has the right to peaceful uses of nuclear energy. The other parties expressed their respect and agreed to discuss, at an appropriate time, the subject of the provision of light water reactor to the DPRK." To Washington, the "appropriate time" is after the North is free of nuclear weapons, if ever, although construction of a reactor without the nuclear components could be started sooner.

A more immediately available alternative would be to provide conventional power as denuclearization proceeds. One possibility is to provide technical and financial assistance for better insulation, more efficient light bulbs, wind power, and other "green energy" initiatives. Other possibilities are to run an electrical line from the South Korean or Chinese grid to special economic zones in the North as was done in the Kaesong Joint Industrial Zone, to provide portable generators, or to construct conventional power plants in the North. In the September 2005 Six-Party Joint Statement, South Korea "reaffirmed its proposal of July 12, 2005 concerning the provision of 2 million kilowatts of electric power to the DPRK." A more demanding alternative is Asian Development Bank aid to refurbish the North's electricity grid.

Political Inducements

Most of the discussion in Washington has focused on economic inducements for North Korean denuclearization. Deepening economic engagement through agricultural and infrastructure aid is a critical element. But it is a misconception to believe that denuclearization can be "bought." It is not for sale. From Pyongyang's point of view, economic engagement is fundamentally political, not economic in nature. It is one aspect of reconciliation.

A key tenet of the North's negotiating position is that denuclearization of the Korean Peninsula will require reciprocal steps by the United States and South Korea to end enmity. Among those steps,

Pyongyang has called for the withdrawal of US troops from Korea, renunciation of the US nuclear threat, and inspections in South Korea to assure that all US nuclear weapons had been withdrawn. US troops remain in Korea at South Korea's behest, so their withdrawal will not be a subject of negotiation with the North without a change of heart in the South.

The US nuclear threat to the North, as the Nuclear Posture Review makes clear, has already receded. A more formal nuclear assurance must await denuclearization by the North. That was explicit in the review's reformulation of the Vance exception, in which the United States forswore the use or the threat of use "against non-nuclear weapons states that are party to the NPT and in compliance with their nuclear non-proliferation obligations." It was also implicit in US pledge in the 1994 Agreed Framework: "The US will provide formal assurances to the DPRK against the threat or use of nuclear weapons by the US" In the September 2005 Joint Statement, the Bush administration stopped short of a formal reassurance: "The United States affirmed that it has no nuclear weapons on the Korean peninsula and has no intention to attack or invade the DPRK with nuclear or conventional weapons." That formal nuclear assurance could be offered as part of a regional security dialogue that addresses the North's security concerns and includes the DPRK as a full participant. Among the subjects for such a dialogue is a nuclear-weapons-free zone for Northeast Asia.[31]

Establishment of diplomatic relations is another potential inducement. That was explicit in the 1994 Agreed Framework, which committed the two sides to "move toward full normalization of political and economic relations." It specifically provided for each side to

---

[31] For a useful version of such a zone, see Morton H. Halperin, "A Proposal for a Nuclear Weapons Free Zone in Northeast Asia," Nautilus Institute, January 3, 2012.

"open a liaison office in the other's capital." Yet North Korea has shown little interest in steps short of full normalization, and the United States has withheld full diplomatic relations as a quid pro quo for complete denuclearization. However, there may be ways to approach this: it may make sense to establish relations sooner in return for steps toward permanent dismantlement. For its part, the United States, short of establishing full diplomatic relations, could offer meetings with the secretary of state and a summit meeting with the president.

Curbing Nuclear Exports

North Korean exports of missiles and missile technology have slowed in the last decade as demand has dropped and the UN-authorized embargo has tightened. Yet North Korea managed to help Syria construct a reactor. In the October 2007 six-party agreement, the DPRK committed "not to transfer nuclear materials, technology, or know-how," but its April 14, 2009 response to UN Security Council sanctions seemed to call that commitment into question. Reaffirmation of that commitment would be useful, but if the North continues to produce fissile material it will eventually have enough to sell, along with improved technology and know-how. That is another compelling reason why permanent dismantlement of its nuclear facilities is worth seeking.

Gradual Reconciliation: The Only Way Out

Will North Korea ever give up its nuclear programs or weapons? Many in Washington and Seoul think not. The answer is to be found in neither of those cities but in Pyongyang. The only way to find it is to offer to move step by step to a fundamentally new relationship with the North as it moves to denuclearize. That requires Washington and Seoul to structure a series of agreements that Pyongyang can accept. It also requires that they live up to their obligations as long as Pyongyang, too, does so.

To some observers in the United States and South Korea, negotiation with North Korea is synonymous with appeasement. To others, any improvement in relations with what they see as a hateful regime is repellent. Political opposition around Washington only compounds the uncertainty of dealing with an obsessively opaque state. No doubt, with North Korea as with the economy, President Obama has been dealt a difficult hand. Yet backing away from the table will only increase insecurity for the United States and its allies. Negotiating in the face of political opposition with no guarantee of success is the alternative to a runaway nuclear North Korea.

Disengagement by the United States and South Korea misses an opportunity to bring about desired internal change in the North. It also leaves the North overly dependent on China. Since the late 1980s, North Korea has sought better relations with the United States, South Korea, and Japan as counterweights to Chinese influence, but Kim Jong-il's August 2011 summit meeting with President Dmitri Medvedev may be a sign that instead the North is reverting to its strategy of playing off Russia against China. That could have profound consequences for the future of Korea and Northeast Asia.

Progress on denuclearization in Korea would ease pressures for further proliferation in Northeast Asia and shore up political support for the global nonproliferation regime. That would be a signal achievement for President Obama's nuclear vision.

Chapter 2

Global Partnerships Against the Spread of Weapons and Materials of Mass Destruction: A Russian Perspective

Alexander G. Savelyev

Since the Treaty on the Nonproliferation of Nuclear Weapons, commonly known as the Nonproliferation Treaty, was opened for signature in 1968 and entered into legal force in 1970, many serious developments have taken place in the field of prevention of uncontrolled proliferation of nuclear weapons, nuclear materials, and weapons technology. One of these developments is the Program of Global Partnership Against the Spread of Weapons and Materials of Mass Destruction, known in short as the Global Partnership Program (GPP).

The program was established in June 2002 during the summit in Kananaskis, Canada, of G-8 states: Canada, France, Germany, Italy, Japan, Russia, the United Kingdom, and the United States. The parties intended to allocate US$20 billion to secure and eliminate the threats of weapons of mass destruction (both nuclear and chemical), mostly focusing on Russia and former USSR republics such as Ukraine. The G-8 leaders agreed to "support specific cooperation, initially in Russia, to address nonproliferation, disarmament, counter-terrorism and nuclear safety issues."[1] Today, 21 countries—one of them the Republic of Korea

---

[1] Group of Eight Leaders, "The G8 Global Partnership Against the Spread of Weapons of Mass Destruction," Kananaskis, Canada, June 27, 2002; http://www.fco.gov.uk/Files/kfile/Art%2002%20gp_stat-en.pdf.

—are members of this program. The main sponsor of the program is the United States, which promised to cover 50 percent of this sum, and in May 2011, US State Department, official Bonnie Jenkins declared that "close to US$19 billion has been allocated for the effort to date on projects inside the one-time Soviet Union, including contributions that have helped fund the disassembly of decommissioned Russian nuclear-powered submarines and scientist engagement programs at the International Science and Technology Center in Moscow."[2] Thus, the global partnership should be considered as part of wider efforts of the United States and the international community in the field of nonproliferation and nuclear-threat reduction.

Before the global partnership was formed, the United States and Russia had made serious efforts to prevent nuclear proliferation in practically all of the former USSR republics—in various spheres: military, political, and technical, and including special operations. For example, Russia and the United States successfully invited Ukraine, Belarus, and Kazakhstan to join the START-1 Treaty, as well as the NPT Treaty, as nonnuclear states. It took about three years of intensive negotiation and nearly six years in all to eliminate all the strategic nuclear weapons deployed in the territories of these new states. In the Ukraine alone, the United States spent about $500 million to this end under the Nunn-Lugar Cooperative Threat Reduction Program.[3] Russia, too, contributed in the field of nonproliferation, managing to withdraw to its territory many tactical nuclear weapons that had been deployed and stored in other republics. Due to some unofficial estimates, by 1992 the USSR had in its inventory 14,500 tactical nuclear warheads, of which only about 60 percent were deployed in what is now Russian territory. This means that after the collapse of the Soviet

---

[2] Daily News on Nuclear, Biological, and Chemical Weapons, Terrorism, and Related Issues. May 31, 2011.

[3] Newsru.com, March 26, 2004.

Union, Russia withdrew nearly 6,000 tactical nuclear warheads from Ukraine, Belorussia, Kazakhstan, Moldova, Turkmenistan, Tajikistan, Kirgizstan, and Uzbekistan.[4]

The GPP, as a large program, not only has advantages but also shortcomings, of course. The latter, however, do not outweigh its overall success and achievements during the ten years of its implementation. There is a long list of complaints from both sides—from donors as well as from recipients. Some of the complaints are accepted by all the participants, and thus the initially announced amount of contribution to the program is usually lower than that originally planned. Sometimes donors do not take into account the necessity to adjust the payments regarding implementation. For example, in the case of chemical disarmament, more funds during the first stages of implementation were needed because new storage plants for chemical weapons had to be constructed. In another example, so-called minor donors often do not want simply to transfer money but demand that their contribution be "visible"—that is, be used for real hardware such as buildings and railroads. Such demands make implementation of the program more complicated. Other problems have included the need for closer coordination among states, difficulties to do with domestic legislation of the recipients of the assistance (including ensuring access to corresponding facilities). In my view, these problems can be overcome, and many of them already have been. The parties are gaining experience, and one can say that the prospects for solving the main problems of possible proliferation as the result of the collapse of the Soviet Union are quite good. However, to put briefly a view that is widely accepted in both Russia and abroad, the most serious obstacle to closer cooperation is lack of trust, and this is so in practically all

---

[4] Military Parity, January 6, 2012. http://www.militaryparitet.com/vp/91/.

spheres of security, including that of nonproliferation issues. This is not just an emotional conclusion.

Politically, Russia cannot accept the idea that it might be viewed as a source of proliferation. Political and military leaders of the state always stress that the Russian nuclear arsenal is under strict control and there is no chance that any terrorist group can have even the slightest access to it. Ten years ago, when the GPP started, Russia really lacked the financial and technical means to use weapons of mass destruction—meaning mostly platforms like-nuclear powered submarines. In the chemical-weapons sphere, Russia was not able to implement all the provisions of the Chemical Weapons Convention and joined it only after it received guarantees from foreign partners for assistance in building facilities to safely liquidate its chemical arsenal. The help was gratefully accepted, but it was never agreed in Russia that the main goal of the program was nonproliferation; it was seen as an effort to prevent accidents with nuclear and chemical materials or as assistance to Russia in implementing its international obligations. In any case, Russia, from the beginning, was an equal partner in the GPP and was not the main problem that forced the G-8 to initiate the partnership.

There exists a general difference between Russia and the United States (and most Western countries) in the list of priorities over threats to national security. Official documents of the two states demonstrate this. In the National Security Strategy of the United States (approved in May 2010), international terrorism and the threat of proliferation of weapons of mass destruction are clearly fixed as the main security problems presently facing the United States, which means that those areas will be the main focus of US efforts in the security sphere.[5]

[5] US National Security Strategy, May 2010; http://www.whitehouse.gov/sites/default/files/rss_viewer/national_security_strategy.pdf.

Russia, in its corresponding official document,[6] accepts that the threat of terrorism and proliferation exists, but sees it as being only among other serious challenges to security, such as the possible unilateral use of force, contradictions between the main players in international policy, and so forth. Russia's Military Doctrine of the Russian Federation is more specific.[7] In the list of military dangers and threats to Russia, proliferation of weapons of mass destruction occupies sixth place; international terrorism is only tenth. In this document, possible NATO expansion and attempts to undermine stability are seen as the most dangerous areas. What this means in practice is that the Russian leadership often looks at US actions throughout the world first from the point of view of possible threat to Russia and Russian interests. Only second does it see US action as an attempt to improve US security.

The clearest example of this is over ballistic-missile defense. Starting from 1983, after US President Ronald Reagan declared the Strategic Defense Initiative (SDI), the Soviet Union and its successor, Russia, have never stopped expressing "deep concern" over US ideas and plans to develop and deploy an effective defensive system that could offer reliable protection against strategic and nonstrategic ballistic missiles. During the second half of the 1980s, when the USSR and the United States made a number of important agreements on nuclear-arms control (INF and the START-1 Treaties), only defense problems remained unsolved. Six years of intensive defense and space talks brought no progress at all. As someone who was a part of those talks, I know that throughout the negotiations there was constant deadlock and complete disagreement on practically all the issues concerning the future of the strategic relations of the parties over antiballistic

---

[6] Russian National Security Strategy up to 2020, approved May 12, 2009; http://www.scrf.gov.ru/documents/99.html.

[7] Military Doctrine of Russian Federation, approved by Presidential Decree N. 146. February 5, 2010; http://www.rg.ru/2010/02/10/doktrina-dok.html.

missile defense.[8]

In the 1990s, the parties continued their efforts to build confidence and predictability in that field. Alongside the START-2 Treaty, they managed to reach an agreement (in 1997) on the definition of strategic and nonstrategic ballistic-missile defense systems (the so-called New York Protocols). But these documents (including START-2) never entered into legal force, mostly because of the position of the United States, which refused to ratify them. During that period, Russia and the other former USSR republics intensively cooperated with the United States and the West in general—since 1992, under the Nunn-Lugar Program. But even this close cooperation could not solve the problem of ballistic-missile defense between the two states. They did, however, act within the framework of the ABM Treaty, signed in 1972 and that had unlimited duration. The treaty (and the protocol to this agreement of 1974) limited the deployment of the components of strategic antiballistic systems to 100 interceptors deployed in one region. These components (ABM radars and launchers) must have fixed land-based mode. The treaty prohibited mobile sea and space-based ABM systems and components. In this sense, the treaty offered predictability and confidence to the parties that strategic stability would not be undermined by unilateral actions of any of them in the field of antiballistic-missile defense. The same kind of the confidence was achieved in other fields of nuclear balance through a number of agreements like the INF Treaty, START-1, and others.

As an instrument and a slogan of its "anti-ABM policy," the Russian leadership took "strategic stability" as its principle—a principle repeated several times in both the National Security Concept and the Military Doctrine of the Russian Federation. It was described as a

---

[8] See A. Savelyev and N. Detinov, *The Big Five: Arms Control Decision-Making in the Soviet Union* (Praeger, Publishers, 1995).

condition for further reductions of nuclear and conventional forces—bilaterally (with the United States) and multilaterally. In these documents Russia expressed its readiness for further deep reduction of nuclear forces, but at the same time it called Russian nuclear forces the instrument (factor) to preserve strategic stability and peace.

The very concept of strategic stability is rather primitive. It is based on the capabilities of each side to deliver "unacceptable damage" under a worst-case scenario—that is, if the aggressor launches a surprise attack and destroys a portion of the strategic system of the opponent. The concept posits that each party to such a scenario possesses relatively "survivable" strategic forces, guaranteeing that under any circumstances the victim of the surprise attack will be able to retaliate. Antiballistic-missile defense plays a very negative role, according to the strategic-stability concept. Such a defense can give a potential aggressor athe illusion that a retaliatory strike could be intercepted. It means that in a crisis, both sides would have serious reason to consider striking first. The possessor of antiballistic-missile defense would be willing to deliver a first strike not only in order to "win" a nuclear war but also because of understanding that the opponent was not sure of the effectiveness of retaliation and would have no choice but to preempt.

The strategic-stability concept was popular in the United States during the Cold War. The USSR, too, agreed that antiballistic-missile defense could play a negative role during that period of central confrontation, and that it could initiate a further arms race, in both offensive and defensive strategic forces. This mutual understanding allowed serious success in arms control. Moreover, the strategic-stability principle was reflected in the INF (1987) and START-1 (1991) treaties. Today, the situation has changed. Russia and the United States have declared that they are not "potential enemies" any more. But suspicions remain, and the Russian leadership still follows this principle, even though it is practically outdated and does not correspond to the realities of the 21st century.

Having understood that security policy based on simple rejection of US proposals and ideas concerning ballistic-missile defense could be counterproductive, Russia's leadership took steps to present a plan alternative to the US one. In June, 2000, Russia's President Putin put forward a proposal to develop together with the European countries a tactical BMD system (which was not prohibited by the ABM Treaty) to provide a reliable defense for Europe and Russia from ballistic-missile threat. The idea was put forward during Putin's "working" visit to Italy. The Russian president asked the Italian prime minister, G. Amato, to be the "mediator" for the proposal and to "investigate" its perspectives with the other European leaders. From tactical viewpoint, this "initiative" was well worked out, and a few days after it was anounced, the Russian defense minister, I. Sergeev, presented its details at a Russia-NATO meeting in Brussels. He offered a list of possible fields of cooperation:

- joint evaluation of the character and scales of missile proliferation and missile threat
- joint development of a concept for a European BMD system, with the stages of its development and deployment
- establishment of a joint missile warning center
- joint military exercises
- joint research and experiments
- joint development of nonstrategic BMD systems
- development and deployment of a nonstrategic BMD system for the protection of UN peace forces and civil populations

Russia also declared that it was ready to have even closer cooperation, but it could be possible only if the ABM Treaty remained in force. What of the "strategic" approach to the problem? Putin's proposal played first a more negative, then a positive role for Russia. By putting it forward, Russia indirectly accepted that the missile threat for Europe could exist, and, moreover, that this threat could be eliminated by the development and deployment of a nonstrategic BMD system. In other words, Russia accepted that ballistic-missile defense could play a positive role from the point of view of security—and, it follows, for strategic stability at a regional level.

The situation radically changed after the tragic events of September 2001. The United States considered that the ABM Treaty contradicted their supreme interests of national security and withdrew from it in 2002 (after a six-month notification, as was provided for in this document). From this moment, the US-Russian strategic relations started to move in two main directions. On the one hand, the parties actively cooperated in arms control and nonproliferation spheres and generally acted as real partners. On the other hand, they failed to find a common understanding on the problem of ballistic-missile defense. While the United States limited its activities in the ABM field to its own territory and deployed defensive systems in the Alaska region, Russia did not react very strongly to this move. But as soon as the United States declared a possible deployment of these systems in Europe, Russia expressed a very negative attitude. Thereafter, Russian suspicions about "the real plans" of the United States grew progressively.

Of course, such a situation did not help either side to achieve progress in global partnership, although the GPP continued to work rather successfully. One of the most important parts of the program deals with utilization of Russia's nuclear-powered submarines. With the assistance of foreign partners, all the infrastructure needed for the submarines' operation was built, mostly in the northern part of Russia. This allows safe use of up to 20 nuclear submarines per year.[9] Many

---

[9] See "Results and Perspectives of the Global Partnership Program," ed. A. Pikaev (Moscow: IMEMO RAN, 2009), p. 41.

countries contributed to the program, but at each stage there were a number of problems that, in my view, were mostly based on a lack of understanding and growing suspicion from both sides about the real aims of the main participants—Russia and the United States. It is obvious that, until the two states find common understanding in their approaches to international security, one cannot expect real progress in global partnership.

The Russian political-military leadership considers that the most serious challenge to the security of the country was presented by the US plan to deploy an ABM system in East Europe. Practically through all of 2007 and the beginning of 2008, this problem was central to the direction of Russian security policy. From the beginning, Russian officials reacted nervously to the US plan to deploy ABM radar in the Czech Republic and put ten interceptors in Poland. First, the Russian leadership unanimously described this system as "an element of US nuclear strategic forces." In other words, the ABM system in Europe is seen in Moscow as an inseparable part of US strategic forces, whose prime goal is to intercept a retaliatory strike or weaken it to an "acceptable level." Stated differently, this system could seriously undermine strategic stability.

Second, Russia continues to insist that the ABM system in Europe is aimed exclusively against Russian strategic forces since neither North Korea nor Iran possesses ballistic missiles capable of reaching Europe. President Putin compared the US ABM "European initiative" with the deployment Pershing nuclear ballistic missiles in Europe in 1983 and compared the situation with the Cuban missile crisis of 1962.

Third, Russia declared that it would undertake effective "asymmetric" countermeasures in order to reduce this threat and to make the strategic situation more stable. One of these measures is to target the elements of the ABM system in Europe with Russian strategic missiles. Alongside this, some experts, and even military officials, including the chief of Russia's general staff, General Yuri Baluevsky, have spoken straightforwardly about the possibility that Russia could withdraw from the INF Treaty as a reaction and a countermeasure to US ABM deployment in Europe.

Russia has never stopped its attempts to reach agreement with the United States and NATO on ballistic-missile defense. During the decade of the 2000s and at the beginning of the 2010s, when a number of proposals were put forward, all Russian attempts to present an alternative plan to the unilateral US deployment failed. A set of discussions and negotiations between the US and Russian defense and foreign ministers also failed. The appointment of a special envoy of the president of Russia for interaction with NATO on missile defense, as well as that of head of the Russian Mission to NATO, Ambassador D. Rogozin, who was given responsibility for solving these problems, was also fruitless.

In November 2010, during a Lisbon Russia-NATO summit, Russia's President D. Medvedev proposed a "sectoral" approach to a joint BMD system for Europe, whereby every country would be responsible for the defense of a particular geographic sector. For example, Russia would defend its territory and that of some of its neighbors in East Europe from missile attacks aimed at any European country. In return, any BMD deployed by NATO would not extend into Russian territory. This approach could remove the fears that NATO's BMD systems could intercept Russia's nuclear-armed, land-based strategic missiles (ICBMs). This proposal, too, proved to be unacceptable to NATO. Several arguments support the Western position. NATO says that its collective defense commitment cannot be delegated to a non-NATO member, and a number of NATO members —notably, the Baltic states and some other Central and East European countries—insist that only NATO has the right and responsibility for their security.

The Russian demand for so-called "written guarantees" that European ballistic missile defense is not "targeted" against Russia also has met with no understanding or enthusiasm in the United States. The problem is that such a demand, as made by Russia, would mean practically a new, full-scale treaty, with the legal obligations of the United States and NATO not to exceed specific numbers of interceptors and radars as well as not to deploy the elements of ballistic-missile defense in certain regions, close to Russian borders. In the Western view, Russia was seeking to directly influence the security posture of Europe. This is absolutely unacceptable for NATO.

In November 2011, the patience of the Russian leadership came to an end. President Medvedev announced a set of countermeasures in response to US plans. He said that Russia would not agree to participate in a program that, in a relatively short period of time (five, six, or maybe eight years) could weaken Russia's potential for deterrence. He also said that Russia was still open for cooperation in the field of ballistic-missile defense; the partners must take into consideration Russian interests, or success would not be possible. Medvedev also said he had ordered the ministry of defense to take the following steps:

– to immediately put the early warning antimissile radar station in Kaliningrad on combat alert
– to strengthen Russia's air and space defenses and consider the enhanced protection of Russia's strategic nuclear deterrent as a priority within this system
– to equip Russia's new strategic missiles with advanced BMD penetration systems and new, highly effective warheads
– to develop adequate, effective and low-cost measures for disabling NATO's BMD systems, including its information and command and control elements

– and, if the above measures proved to be ineffective, to prepare deployment of weapons in Russia's west and south, such as Iskander missiles in the Kaliningrad Region, capable of eliminating any BMD system in Europe.

Medvedev also said that if the situation developed unfavorably for Russia, Moscow reserved the right to discontinue further disarmament and arms-control measures and, if necessary, exercise the withdrawal clause from the New START agreement.[10]

Many Russian observers commented that in the main Medvedev had simply made public the programs that were in any case planned to happen. Only the last point—the possible deployment of missiles in the Kaliningrad area—could be viewed as counteraction to the deployment of a BMD system in Europe. The same can be said about possible withdrawal of Russia from the new START. One can speculate a lot about Medvedev's "initiative." For example, did the Russian president really believe that he could frighten the West by this set of countermeasures? Why did he set this up for his successor as president of Russia (by this time, Medvedev had officially refused to run for a second term)? Did Medvedev really believe that nonnuclear defensive systems were more dangerous than nuclear offensive forces? Who was behind these "initiatives" and who would profit from it? The list of such questions could be longer, but regardless of the answers it is obvious that the increased tension between Russia and the United States and NATO in general will not contribute to the cooperation in many other fields, including the global partnership, although at an official level Russia has never declared that this program can fall victim to its disagreements with NATO on ballistic-missile defense. All this does not improve trust and predictability among the members of the program.

---

[10] http://www.rg.ru/2011/11/23/prezident.html.

An example of diminishing trust can be seen in the case of the International Science and Technology Center (ISTC). On November 27, 1992, Russia, the United States, Japan, and the European Union had signed an agreement on the establishment of the center. The main objectives of this program were:

- to provide Russian and CIS former weapons scientists, particularly those with knowledge and skills related to weapons of mass destruction and their delivery systems, opportunities to redirect their talents to peaceful activities
- to contribute to solving national and global technological problems
- to support basic and applied research
- to support the transition to market-based economies
- to encourage the integration of Russian and CIS former weapons scientists into the international scientific community

In May 1998, the Republic of Korea became a member of the ISTC,[11] which in 2002 became a part of the GPP. The funded projects of ISTC are:

- Space, aircraft, and surface transportation
- Fission reactors
- Physics
- Biotechnology and life sciences
- Chemistry
- Environmental science
- Nuclear fusion
- Information and communications
- Instrumentation
- Manufacturing technology
- Materials science
- Nonnuclear energy
- Other basic sciences

---

[11] http://www.istc.ru/.

Initially, nearly US$1 billion was allocated to the center. Nearly 3,000 projects were implemented, with 20,000 employees annually at the hundreds of institutes, scientific centers, and other organizations.[12] From the point of view of number of start-ups, the program looks very effective and successful. At the same time, the ISTC is accompanied by growing dissatisfaction and suspicion on the Russian side. In many cases, the results of the activities of Russian scientists can be used by foreign companies, which are associated with their nation's military-industrial complex. This means that foreign states through the ISTC have access to sensitive Russian technologies and can use Russian technological achievements in the interests of a military build-up. This access is absolutely legal, and in practice it means that Russian researchers from the defense industry have started to work for foreign states, many of which are not considered by the Russian authorities to be "friends." Growing dissatisfaction over this was reflected in the Russian legislative body: the State Duma did not ratify the ISTC agreement throughout the period of its implementation (i.e., the ISTC functions in Russia on the basis of a presidential decree and does not enjoy full legal rights in Russia).

All this resulted in a diplomatic note from Russia's ministry of foreign affairs on July 13, 2011, announcing the withdrawal of Russia from the ISTC by 2015. Although the note stated that all the provisions of the ISTC agreement would remain fully in force and all approved projects in the Russian Federation would be completed within the remaining timeframe, it is obvious that such a move will not contribute to the global partnership, and general trust between the partners of the GPP. In this situation, the parties propose to discuss the possible establishment of a new multilateral network for scientific,

---

[12] *The Results and Perspectives of the Global Partnership Program*, ed. A. Pikaev (Moscow: IMEMO RAN, 2009), p. 99.

technological, and nonproliferation cooperation.[13] But clearly, the present format does not satisfy all the participants, notably Russia. The main reason for it, in my view, is the lack of trust caused by other international military and political factors, including the European BMD system discussed above.

In spite of such problems, the GPP has many more advantages than shortcomings. The most successful and fruitful field of its application is cooperation with the Russian State Corporation on Atomic Energy (Rosatom). Substantial results have been achieved in this area, among them:

- Reduction of the nuclear-submarine-dismantling period by at least three years
- Construction and commissioning of a radioactive-waste long-term storage facility in the Northwest of Russia (Saida Bay), where 20 reactor compartment units have been placed in storage
- Construction and commissioning of physical protection units of SevRAO company radiation-hazard facilities
- Commissioning of a nonreprocessible SNF (spent nuclear fuel) storage facility has been commissioned at the Murmansk Shipping Company RTP Atomflot
- Establishment and operation of an emergency-response system in the Murmansk region
- Construction of facilities to support safe SNF unloading and nuclear-submarine dismantling
- Determination of safe technologies for SNF management at Andreeva Bay and a start made on support-unit construction
- Development of a strategic master plan for determining the long-term prospects of the Russian Northwest.

Up to 2012, the priorities have included the completion of nuclear-submarine dismantling; the decommissioning of surface ships with nuclear-power installations; the completion of a radioactive-waste,

---

[13] http://www.istc.ru/.

long-term storage facility at Saida Bay; preparation for SNF removal from interim storage facilities at Andreeva Bay and Gremikha.

A number of programs are planned for after 2012: dismantling surface ships with nuclear-power installations, nuclear icebreakers, and nuclear maintenance vessels; isolation of nuclear submarines with damaged reactor facilities; supply of equipment and completion of construction of an onshore reactor-compartment storage facility in the Far East; provision of SNF removal for reprocessing from storage sites at Sysoeva Bay, Andreeva Bay, and Gremikha; construction of a regional radioactive-waste conditioning and long-term storage center in the Russian Northwest and Far East; placing reactor units for storage; project development and remediation of radiation hazardous facilities in the North-West Region and in the Far East; and others.[14]

Analysis of information on the GPP show that, in the various sectors of implementation, the different parties see different results. Nevertheless the program in general can be described as successful. During a ten-year period of cooperation the states have gained experience—positive as well as negative. In my view, the positive results outweigh all the shortcomings and problems.

In terms of increasing the efficiency of international cooperation, the factors involved can be combined into several categories, among them financial, technical, legal, political. In each case, the parties had to solve many problems, relatively new for them, ranging from confrontation to cooperation. It was not only a matter of changing the mentality of the participants; it also involved differences in technical cultures, in contract systems, in understanding the meaning of *cooperation*, and in experience in dealing with foreign partners, which included different levels of involvement of private and state-owned enterprises and

---

[14] E. G. Kudryavtsev and A. V. Grigoriev, Main Results of Global Partnership Program Implementation in Russia, Paris, September 24, 2008.

companies. In most cases, the participants managed to adjust to each other and establish effective international structures.

From personal experience, the author found that there is a specific rule in any international program: it is much easier to achieve understanding at the lower level of executives than at the higher levels of bureaucracy. In case of the GPP, the highest level is political, and decisions taken at this level can either accelerate the program or slow it down. The political level can play a very positive role in international cooperation and can help to solve serious problems by taking proper decisions. The partnership itself is the product of political decisions, and there is no reason, in my view, to stop it.

The process of cooperation starts to influence policy—starts to present a real alternative to confrontation and misperception. And the more successful cooperation there is, the more positive results it brings toward improvement of political relations between states. The GPP is a very good example of both the negative and positive trends, with the positive outweighing the negative. In my view, the GPP must be continued. The alternative would increase misunderstanding and suspicion.

Chapter 3

Reassessing the G-8 Global Partnership: A South Korean Perspective

Sang-Hyun Lee

Nuclear-weapons proliferation, whether by state or nonstate actors, poses one of the greatest threats to international security today. Iran's apparent efforts to acquire nuclear weapons, what amounts to North Korean nuclear blackmail, and the revelation of the A. Q. Khan black-market nuclear network all underscore the far-from-remote possibility that a terrorist group or a so-called rogue state will acquire weapons of mass destruction (WMD) or materials for a dirty bomb. In April 2009, US President Barack Obama reignited global nonproliferation efforts through a landmark speech in Prague and subsequent actions on the international stage.

The problem of nuclear proliferation is global, and any effective response must be multilateral. Nine states—China, France, India, Israel, North Korea, Pakistan, Russia, the United Kingdom, and the United States—are known or believed to have nuclear weapons, and more than thirty states are believed to have the technological ability to quickly acquire them.[1]

In the post–September 11 world, prevention of further nuclear

---

[1] "The Global Nuclear Nonproliferation Regime," backgrounder, Council on Foreign Relations, updated January 9, 2012; http://www.cfr.org/proliferation/global-nuclear-nonproliferation-regime/p18984.

proliferation by both state and nonstate actors became even more important than before. The September 11 terrorist attacks in 2001 were a wake-up call that alarmed the international community, alerting it to the looming danger of nuclear terrorism and necessitating a quick reaction. However, shortages of resources, high-level attention, international coordination, and global norms have limited their effects.

South Korea hosted the Nuclear Security Summit on March 26 and 27, 2012 in Seoul. The event provided a global arena in which the leaders of the international community could discuss how to secure the world from dangerous weapons and materials. In this chapter, the author reviews the historical background of the G-8 Global Partnership and its major achievements and suggests policy recommendations to further strengthen the global nuclear nonproliferation regime.

The G-8 Global Partnership and Global Nuclear-Nonproliferation Regimes

According to a 2010 International Panel of Fissile Materials (IPFM) report, there are more than 2,100 tons of fissile material stock in the world—1,600 tons of highly enriched uranium (HEU) and about 500 tons of separated plutonium—enough to produce more than 100,000 nuclear weapons.[2] A large stock of HEU for civilian use is a cause for special concern since it is under far less protection than the material for military use, and the gun-type HEU bomb requires fewer engineering skills in assembly and explosion than the plutonium-core explosion bomb. The IAEA Illicit Trafficking Database also shows that nuclear, radiological terrorism can occur. There have been more than 20 cases

---

[2] "Global Fissile Material Report, 2010: Nuclear Weapon and Fissile Material Stockpiles and Production," International Panel on Fissile Materials, 2011; http://www.ipfmlibrary.org/gfmr10.pdf.

of illicit use or trade of HEU and plutonium from1993 to 2011.[3]

Despite the overwhelming danger of intended or unintended misuse of nuclear materials, the international laws on nuclear security are still at an early stage of formation, and in most cases lack binding force. The only two international treaties on nuclear security with binding force are the Convention on the Physical Protection of Nuclear Material (CPPNM) and the International Convention for the Suppression of Acts of Nuclear Terrorism (ICSANT). The CPPNM, which came into effect in February 1987, is the only legally binding international instrument dealing with physical protection of nuclear material and other radiological sources. It requires national legislation to prevent and punish nuclear-material-related offenses. The initial CPPNM covered only international transportation of nuclear material, but the 2005 Amendment called for the protection of nuclear material in peaceful domestic use, storage, and transport as well as at nuclear facilities. This new amendment also asks for states to respond quickly to locate and recover stolen or smuggled nuclear material, to mitigate any radiological consequences of sabotage, and to prevent and combat related offenses. Since the 2005 Amendment is not yet in effect, lacking ratification by the required two-thirds of member states, its early effectuation has become one of the primary goals of the Nuclear Security Summit.

The ICSANT, based on the draft first proposed by Russia in 1998, was adopted by the United Nations General Assembly and opened for signing in 2005. This is a key element of global efforts to prevent terrorists from gaining access to weapons of mass destruction. The convention provides for a definition of acts of nuclear terrorism and covers a broad range of offenses, including those against nuclear-

---

3 Illicit Trafficking Database, International Atomic Energy Agency; http://www-ns.iaea.org/security/itdb.asp.

power plants and nuclear reactors. The ICSANT also obligates states to extradite or prosecute alleged offenders and encourages them to share information and assist each other during criminal investigations and extradition proceedings.

In addition, there has been promising progress in international organizations regarding the nonproliferation of nuclear materials and technologies. The United Nations Security Council adopted Resolution 1540 in 2004 under Chapter VII of the UN Charter. This resolution binds UN member states to stop supporting non-state actors seeking weapons of mass destruction, to adopt and enforce effective laws prohibiting the proliferation of such items to non-state actors and their financing, and to enforce effective measures for nonproliferation. Though the UNSCR 1540 Committee was established to accelerate implementation of the resolution, its mandates and resources have been insufficient to fulfill its goals of nonproliferation and nuclear security.

Along with UN activities, the International Atomic Energy Agency (IAEA) has expanded its function in order to adapt to a new international security environment after the 9/11 incident. First, the IAEA's Information Circular (INFCIRC) 225, which provides guidance and recommendations for developing and implementing the physical protection of nuclear material and nuclear facilities, was revised for the fifth time in 2011 and published as INFCIRC/225/Rev. 5. Second, the IAEA also published the IAEA Nuclear Security Plan 2010–2013 and the IAEA Nuclear Security Series to provide guidelines for nuclear security. Finally, the International Physical Protection Advisory Service (IPPAS) was created to assist its member states in implementing various nuclear-security instruments and guidelines.[4]

---

[4] Bong-Geun Jun, "Road to the 2012 Seoul Nuclear Security Summit," US-Korea Institute at SAIS, February 2012; http://uskoreainstitute.org/wp-content/uploads/2012/02/USKI_NSS2012_JUN.pdf.

Other mechanisms are particularly focused on preventing nuclear terrorism and the spread of WMD and related materials. The Global Initiative to Combat Nuclear Terrorism (GICNT), proposed and cochaired by the United States and Russia, was launched in 2006 to develop a partnership to combat nuclear terrorism. The GICNT is a voluntary cooperative framework that, as of 2011, has been joined by 82 members and four observers. Members have been asked on a "voluntary basis" to implement principles such as developing accounting, control, and physical protection systems for nuclear and radioactive materials; enhancing the security of civilian nuclear facilities; improving detection capabilities to prevent the illicit trafficking of such materials; and supporting the IAEA's nuclear security functions.

Another successful example of global efforts to curb the spread of WMD is the Proliferation Security Initiative (PSI). The US-led PSI, established in 2003, today involves more than ninety countries in developing best practices, joint training exercises, and information-sharing activities to improve multilateral interdiction efforts. Although often cited as a flexible approach to coordinating the international response to proliferation, PSI does not grant any legal authority for ship-boarding or interdiction beyond the UN Law of the Sea treaty and various bilateral agreements. Additionally, India and China, which do not participate in PSI, have questioned the legality of its interdictions. PSI also cannot interdict ships of nonmember states unless the master consents to being boarded.[5] Whether the 2003 interdiction of a ship supplying nuclear materials to Libya was the direct result of PSI activities, for example, is still disputed. Analysts have also criticized the PSI for being a club of developed economies and not addressing the problem of increasing independence among a

---

[5] For further detail regarding the PSI, see US State Department website; http://www.state.gov/t/isn/c27732.htm.

growing number of developing countries and nonstate actors from the controls enacted by the traditional supporters of the nuclear establishment. Others have pointed out that the initiative is limited by having neither an independent budget nor coordinating mechanisms and does not provide a legal framework in which to lock-in long-term, verifiable, and irreversible member-state commitments.

All these efforts well illustrate the past progress and achievements in the area of nuclear nonproliferation. Although successful, these are not enough. In dealing with today's proliferation challenges, international organizations work in tandem with ad hoc forums of interested parties, such as the Six Party Talks on North Korea and the P5+1 grouping on Iran. But such forums have often proven inadequate to arrest the spread of nuclear technology, and states such as Iran and North Korea continue to pursue nuclear capability, if not outright weaponization. Given these trends, rising doubts about the sustainability of the nonproliferation regime are no surprise. Furthermore, major weakness in the global nonproliferation regime includes: incomplete coverage, inadequate monitoring and enforcement mechanisms, gaps in the legal regime, inadequate regulation of dual-use technologies, failure of the nuclear-weapon states (NWS) to meet disarmament commitments, and lack of legitimacy (nonnuclear weapon states [NNWS] often do not fully accept the NPT, UNSCR 1540, and other key elements of the nonproliferation regime).

There are also more cooperative approaches. Possibly the most successful element of the nonproliferation regime has been the effort to secure so-called loose nukes and fissile material throughout the former Soviet Union. This is critical given that some 135 nuclear facilities worldwide use highly enriched uranium (HEU) as fuel—enough HEU to create some 400 nuclear weapons. If terrorist or criminal groups were able to buy or steal even a small portion of this material, they could use it to construct a crude nuclear weapon or dirty bomb.

US Senators Sam Nunn and Richard Lugar sponsored the initiative that became Title 22 (chapter 68a) of US Code and authorized the Cooperative Threat Reduction program—also called the Nunn-Lugar Program. The program initially intended to prevent the proliferation of weapons of mass destruction and related materials, technologies, and expertise from former Soviet Union states. This included providing for the safe destruction of Soviet era WMD, associated delivery systems and related infrastructure, and consolidating and securing former Soviet states' WMD and related technology and materials. The US Congress established the CTR Program in December 1991 as the Soviet Nuclear Threat Reduction Act—later renamed as the Cooperative Threat Reduction Program in 1993 legislation and has received more than $5.9 billion through annual defense appropriations since 1991. Assistance has been provided to the former Soviet republics—primarily Russia, Ukraine, Kazakhstan, Georgia, Uzbekistan, and Azerbaijan.

The Global Partnership (hereafter, GP) against the Spread of Weapons and Materials of Mass Destruction represents such a cooperative approach. The G-8 GP began at the 2002 Kananaskis G-8 Summit as a ten-year, $20 billion initiative to prevent terrorists or states that support them from acquiring or developing weapons of mass destruction (WMD). Since then, the GP has grown to include 23 partners and has allocated about $21 billion worldwide. At the 2011 G8 Summit in Deauville, leaders agreed to extend the partnership beyond 2012. The global partnership addresses nonproliferation, disarmament, counterterrorism, and nuclear-safety issues through cooperative projects in such areas as destruction of chemical weapons; the dismantlement of decommissioned nuclear submarines; the security and disposition of fissile materials; and rechanneling employment of former weapons scientists to peaceful civilian endeavors.

Its key achievements to date include:

- Improved accounting, control, and physical protection of nuclear and radiological materials
- Enhanced nuclear, biological, and chemical security
- Dismantled nuclear submarines and safe storage of removed spent fuel
- Improved detection of nuclear and radiological materials and prevention of illicit trafficking by improved border-security capabilities
- Engaged and redirected to peaceful purposes scientists, technicians, and engineers who have WMD, missile, and related expertise
- Provided enhanced training on nuclear safeguards and security

The global partnership is the primary multilateral arrangement for financial commitments to implement and coordinate chemical, biological, radiological, and nuclear-threat reduction activities on a global scale. Comprising 23 partners, the GP is committed to preventing proliferators, terrorists, or other nonstate actors from acquiring weapons of mass destruction. A partnership working group meets four to five times a year to engage in a dialogue regarding the progress of initiatives supported by this key multilateral forum.

The Global Partnership also produces an annual annex that details each nation's projects under the GP. Because the current ten-year, $20 billion GP commitments will expire in 2012, it is time to extend the GP so it can continue building upon its successful efforts to combat the spread of WMD, their delivery systems, and related technology. Canada has proposed extending the partnership for its 2010 G-8 presidency, and the United States strongly supports that proposal.[6] The G-8 GP is a unique case that intends to strengthen the global nuclear-nonproliferation regime as a cooperative endeavor.

---

[6] Bonnie D. Jenkins, "The Future Role of the G-8 Global Partnership: Combating Weapons of Mass Destruction," Stanley Foundation, *Policy Analysis Brief*, June 2010; http://www.state.gov/documents/organization/184789.pdf.

South Korea joined the GP in June 2004 and pledged to donate 5.5 million US dollars during 2005 to 2011. The following table summarizes South Korea's record on the GP:

Table 1. **South Korea's Contribution to G-8 GP, 2005–2011**

Year	Amount	Key Projects
2005	$500,000	– Construction of a thermal power plant in Zheleznogorsk, Russia (joint project with the United States) – Decommissioning Russian retired nuclear submarines (joint project with Norway)
2006	$500,000	– Construction of a thermal power plant in Zheleznogorsk, Russia (joint project with the United States) – Decommissioning Russian retired nuclear submarines (joint project with Norway)
2007	$1,000,000	– Decommissioning retired Russian nuclear submaries (joint project of South Korea and Japan) – Canadian-led project for enhancing physical protection of Russian nuclear facilities – Construction of a thermal power plant in Zheleznogorsk, Russia (with the United States) – US-led Nuclear Smuggling Outreach Initiative (NSOI) program in Ukraine
2008	$1,000,000	– Canadian-led program for decommissioning retired Russian nuclear submarines – Japanese-led program for decommissioning retired Russian nuclear submarines – US-led NSOI program
2009	$700,000	– Canadian-led program for decommissioning retired Russian nuclear submarines – US-led NSOI program in Kazakhstan – US-led program for enhancing Biosecurity Engagement Program (BEP) in Afghanistan
2010	$900,000	– Canadian-led program for decommissioning retired Russian nuclear submarines – US-led NSOI program in Ukraine – US-led program for enhancing Biosecurity Engagement Program (BEP) in Afghanistan – Partnership for Nuclear Security (PNS) in the Middle East

Table 1. **South Korea's Contribution to G-8 GP, 2005–2011 (Continued)**

Year	Amount	Key Projects
2011	$900,000	– US-led NSOI program in Ukraine and Kazakhstan – US-led program for enhancing Biosecurity Engagement Program (BEP) in Afghanistan – Partnership for Nuclear Security (PNS) in the Middle East

In addition to the above contributions, South Korea has assisted the operations of the International Science and Technology Center (ISTC) in Moscow with funds from South Korea's Ministry of Education, Science, and Technology.

The Seoul Nuclear Summit and Beyond: Toward a Synergy Between Security, Safety, and Safeguards

Fifty-three heads of state and government, as well as representatives of the United Nations (UN), International Atomic Energy Agency (IAEA), European Union (EU) and INTERPOL, attended the 2012 Seoul Nuclear Security Summit.[7] Compared to the 2010 Washington Summit, there were seven new participants: Azerbaijan, Denmark, Gabon, Hungary, Lithuania, Romania and INTERPOL. The EU was represented by both the President of the European Council and the President of the European Commission, making the number of partic-

[7] Republic of Korea (Chair), Algeria, Argentina, Armenia, Australia, Azerbaijan, Belgium, Brazil, Canada, Chile, China, Czech Republic, Denmark, Egypt, Finland, France, Gabon, Georgia, Germany, Hungary, India, Indonesia, Israel, Italy, Japan, Jordan, Kazakhstan, Lithuania, Malaysia, Mexico, Morocco, The Netherlands, New Zealand, Nigeria, Norway, Pakistan, Philippines, Poland, Romania, Russia, Saudi Arabia, Singapore, South Africa, Spain, Sweden, Switzerland, Thailand, Turkey, United Arab Emirates, Ukraine, United Kingdom, United States, and Viet Nam.

ipating leaders 58 in total.

The Seoul Summit was held on March 26–27, 2012 at the COEX Centre. The Seoul Communiqué, which is the official document of the Seoul summit, built on the objectives and measures set out in the 2010 Washington Communiqué to identify 11 areas of priority and importance in nuclear security and presents specific actions in each area. The 11 areas are as follows: the global nuclear security architecture; the role of the IAEA; nuclear materials; radioactive sources; nuclear security and safety; transportation security; combating illicit trafficking; nuclear forensics; nuclear security culture; information security; and international cooperation.

The Seoul Communiqué sets out the following specific actions in the above 11 areas:[8]

- Eliminating and disposing of highly enriched uranium (HEU) no longer in use
- Minimizing the use of HEU
- Encouraging voluntary announcements by the end of 2013 of specific actions to minimize the use of HEU
- Welcoming international efforts to develop high-density low-enriched uranium (LEU) fuel for the purpose of replacing HEU fuels in research reactors and medical isotope production facilities
- Seeking to bring the 2005 amended Convention on the Physical Protection of Nuclear Materials (CPPNM) into effect by 2014
- Welcoming an international conference in 2013 organized by the IAEA to coordinate nuclear security activities
- Encouraging voluntary contributions to the IAEA Nuclear Security Fund
- Developing options for national policies on HEU management within the framework of the IAEA

---

[8] "2012 Seoul Nuclear Security Summit: Key Facts," Ministry of Foreign Affairs and Trade (http://gab.mofat.go.kr/webmodule/htsboard/template/read/legengreadboard.jsp?boardid=12982&c=&dc=&du=&iu=&lu=&pagenum=1&pc=&seqno=675227&t=&tableName=TYPE_ENGLEGATIO&typeID=16&vu=&wc=, posted on: April 5, 2012).

- Encouraging national measures and international cooperation to prevent radiological terrorism
- Strengthening the physical protection of nuclear facilities and enhancing emergency response capabilities in the case of radiological accidents while comprehensively addressing nuclear security and nuclear safety concerns
- Strengthening the management of spent nuclear fuels and radioactive wastes
- Strengthening the protection of nuclear materials and radioactive sources in transport
- Encouraging the establishment of a system to effectively manage and track such materials on a national level
- Preventing the illicit trafficking of nuclear materials
- Strengthening technical capabilities to search for and detect illicitly trafficked nuclear materials and encouraging the sharing of information on persons involved in such activities by cooperating with the INTERPOL
- Building nuclear forensics capacity to identify the source of illicitly trafficked nuclear materials
- Welcoming the establishment of Centers of Excellence for training and education in nuclear security, and supporting networking activities between each Center
- Strengthening the nuclear security culture
- Encouraging the participation of industry, academia, the media, NGOs and other civil actors in the discussions on nuclear security
- Strengthening the protection of sensitive nuclear security-related information and enhancing cyber security at nuclear facilities
- Promoting international cooperation, such as the provision of assistance to countries for the enhancement of national nuclear security capabilities upon request
- The hosting of the next Nuclear Security Summit in the Netherlands

There are a number of points particularly worthy of note in the Seoul Communiqué. Firstly, it provided important timelines for advancing nuclear security objectives, such as the target year (end of 2013) for states to announce voluntary actions on minimizing the use of HEU and the goal year (2014) for bringing the amended CPPNM into effect. Secondly, it reflected the need to address both the issues of nuclear security and nuclear safety in a coherent manner for the sustainable peaceful uses of nuclear energy. It also emphasized the need to better secure spent nuclear fuel and radioactive waste. Thirdly, it set out specific measures to prevent radiological terrorism, an issue which was only briefly touched upon at the Washington Summit.

During the Seoul summit, 32 countries made over 70 commitments on specific actions to enhance nuclear security at the Washington Summit, and the national progress reports submitted by the participating countries have shown that nearly all of these have been achieved. Likewise, over 100 commitments were made from participating countries at the Seoul Summit.

Other than these, there are numerous accomplishments achieved during the Seoul summit. The Seoul summit made a substantial progress on the issue areas such as removing HEU or converting HEU to non-military use, disposing and security plutonium, converting research reactors and medical isotope production facilities using HEU fuel to LEU fuel, strengthening nuclear security-related international conventions and multilateral initiatives, establishing Centers of Excellence, supporting the activities of the IAEA, countering the illicit trafficking of nuclear and radiological materials, hosting of nuclear security conferences and events, and so on. It was decided that the next Nuclear Security Summit will be held in 2014 in the Netherlands. Several sherpa meetings and sous-sherpa meetings will be held in the lead up to the Netherlands Summit.

From South Korea's perspective, it is important to look beyond the Seoul summit so that its momentum can be carried over to the next round of the summit in 2014. At the same time, it is imperative to pursue a way for continuing the legacies of the two nuclear-security summits. It's time to think about the post-NSS world.

In order to guarantee full function, an international nuclear nonproliferation regime should be comprehensive—be universally applicable. It also requires that the international community should work toward a synergy effect between the GP and nuclear-security summits. Nuclear safety and security have developed along different trajectories in past decades. The regime that supports nuclear safety—a combination of national laws and regulations, voluntary international agreements, and conventions—matured quickly following the accidents at Three Mile Island and Chernobyl. The nuclear-security regime, on the other hand, has advanced largely in response to the terrorist attacks on September 11, 2001, in the United States. In the post-Fukushima world, it is essential to view the nonproliferation issue in the integrated framework of nuclear security, safety, and safeguards.

At the 2008 Hokkaido summit in Japan, and again at the L'Aquila summit in 2009, the G-8 leaders agreed to expand membership of the GP. This was consistent with the principles established at Kananaskis in 2002, setting up the structure for the work of the GP. In addition to expanding geographically, the GP must also grow in size. To meet the global threat effectively, the GP must expand to include new members and, in particular, look to add potential regional leaders. Many nations could be considered in such an expansion, especially those that attended the two previous Nuclear Security Summits. Quite naturally some experts argue that the GP should develop as a successor forum to the summit. The GP has aroused suspicions in nations of the South, given that it is a creature of the G-8 and therefore of the Western powers. However, the GP is on course to be substantially expanded and reoriented—which may offer a chance for developing nations to join with a greater sense of ownership.

It is also recommended to place the NSS under the aegis of the G-20. An advantage of the G-20 is that it is focused on consensus building. And even though the group does not have universal membership, its members possess more than 90 percent of the world's nuclear weapons, constitute more than 70 percent of the world's gross domestic product, and are home to 80 percent of the world's population. Such a concentration would make resistance from outside states more difficult, and would likely bolster the norm-building process. However, this idea drew the oft-heard response that the proper focus of the G-20 is economic affairs. Even participants who support G-20 moves into political matters saw problems with the mismatch of the two forums' participating states and the adoption by the G-20 of such an extensive program. (Advocates of a broadened G-20 agenda have focused more on issues to be upscaled, rather than downshifted, as would be with the NSS.) The question was also raised as to whether this was the right group of countries to address nuclear security.[9]

When delegates from 49 nations convened on January 16–17, 2012, in New Delhi for the Sherpa meeting to work on the agenda for the Seoul summit, Indian Foreign Secretary Ranjan Mathai said that "the main objective of the nuclear summit process has been to focus high-level global attention on the threat posed by nuclear terrorism." He further said that "security of nuclear materials is fundamentally a national responsibility but there is a considerable scope for international cooperation to strengthen nuclear security objectives and standards." Indeed, security of nuclear materials is fundamentally a national responsibility, but there is considerable scope for international cooperation to strengthen nuclear security as well as combating illicit trafficking. The preparatory meeting of the forum held in New Delhi was led by the United States and South Korea. It focused principally on reaching

---

[9] "Nuclear and WMD Security and Summit Diplomacy: Leveraging Top-Level Engagement," Stanley Foundation, Policy Dialogue Brief, A Conference Report of the 52nd Strategy for Peace Conference sponsored by the Stanley Foundation, October 13–15, 2011, Warrenton, Virginia; http://www.state.gov/documents/organization/185073.pdf, pp. 4–5.

agreement on a preliminary document that would be approved by participating nations at the Seoul summit.[10]

One of the central issues covered at the event in Seoul was how to safeguard radioactive sources around the world from potential diversion by extremists seeking to produce an improved nuclear weapon. Additional agenda issues included practical and concrete options for minimizing the danger of a nuclear terrorist attack and preventing nuclear-weapon crises like that at Japan's Fukushima Daiichi facility.

US nuclear expert Kenneth Luongo testified before the US House of Representatives Committee on Foreign Affairs shortly after the 2010 Washington summit, arguing that the international community should create a global nuclear-security "gold" standard that all nations could accept. Luongo pointed out that "despite the detailed technical information that is provided by the IAEA for the safeguarding of nuclear facilities and the other domestic and international conventions and regulations that govern nuclear material protection, no universally accepted standard exists for securing nuclear materials and weapons."[11] The Seoul summit was a landmark test of the willingness of the international community toward such a "gold standard."

A final controversial issue at the Seoul Summit was the relation between GP, the summit, and North Korea's continuing nuclear saga. As the host country, South Korea was concerned that the summit should have implications for North Korea, though the international gathering was not expected to specifically take up the issue of Pyongyang's nuclear-weapons development. Though deterring terrorists from acquiring and using nuclear materials was the main objective of the

---

[10] Rajaram Panda, "2012 Global Nuclear Security Summit in Seoul: Views from India," *JPI PeaceNet*, March 1, 2012.

[11] Kenneth Luongo, "The Nuclear Security Summit Achievements and Agenda for Action," testimony before the United States House of Representatives Committee on Foreign Affairs, April 21, 2010: 10; http://foreignaffairs.house.gov/111/luo042110.pdf.

Seoul summit, South Korea hoped to send a message to North Korea in one way or another. The Stalinist state is viewed as a top proliferator of missiles and other WMD-related materials and technologies and is suspected of conducting nuclear collaboration with nations such as Syria. A majority of North Korea experts in Seoul do not expect the new North Korean leader Kim Jong-un to substantially alter his father's policies on nuclear weapons. Kim Jong-il, who died in December 2011, saw nuclear arms as his regime's insurance policy against a possible invasion by the United States and as a crucial tool for securing economic and security concessions from the international community.

In November 2010, North Korea had revealed a new nuclear facility to a visiting delegation of US nuclear scientists. The previously secret facility contains 2,000 steel-rotor centrifuges producing low-enriched uranium—a material that can be used to make nuclear weapons. Although North Korea already possesses a small nuclear arsenal, its previous nuclear weapons were made from plutonium. Pyongyang, however, insists that the uranium-enrichment plant is intended to power a light-water reactor to generate civilian nuclear technology.

During a trip to Russia in August 2011, however, North Korean leader Kim Jong-il repeatedly expressed his readiness to return to the Six Party Talks in exchange for international aid and other concessions. The Six Party Talks, aimed at dismantling the North Korean nuclear program, stalled in 2008 when North Korea walked out of the negotiations over verification measures for nuclear disarmament. North Korea's nominal willingness to resume negotiations was prompted by an urgent need for economic aid to alleviate devastating food shortages.

The sudden death of North Korean leader Kim Jong-Il in late December 2011 added fresh concern about the safety of North Korea's nuclear arsenal. US policymakers and other nuclear experts then monitored North Korea carefully since it appeared to be under leadership transition from Kim Jong-il to Kim Jong-un.

The latest twist in North Korea's nuclear saga is its third nuclear test. North Korea conducted its third nuclear test on February 12, 2013 in defiance of previous U.N. Security Council resolutions, drawing condemnation from around the world, including from its only major ally, China, which summoned the North Korean ambassador to protest. Pyongyang said the test was an act of self-defense against 'U.S. hostility' and threatened stronger steps if necessary. The test put pressure on U.S. President Barack Obama on the day of his State of the Union speech and also put China in a tight spot, since it came in defiance of Beijing's admonishments to North Korea to avoid escalating tensions.

The implication of the North Korea nuclear issue is straightforward. North Korea's nuclear-development program poses a serious threat not only to the regional security situation on and around the Korean peninsula but also to cumulated achievements in the area of global nuclear nonproliferation. The international community must pay higher attention to the issue for an earlier, rather than later, resolution. The nations concerned should work together to induce North Korea onto the road toward complete denuclearization. Some experts argue that, rather than wait for signs out of Pyongyang, the United States should now signal its interest in developing a more productive relationship. We don't know whether there are pragmatic and potentially reform-minded elements in the leadership, but we should at least try to find out and even encourage them. Such signaling should include offers of food and other humanitarian aid to alleviate widespread shortages, as well as a willingness to engage in renewed efforts to negotiate limits on North Korea's nuclear-weapons and missile programs.[12]

---

[12] Paul B. Stares, "Seize the Moment in North Korea," *New York Times*, December 22, 2011; http://www.nytimes.com/2011/12/23/opinion/seize-the-moment-in-north-korea.html?_r=2&ref=global.

A joint effort to foster stable change in North Korea will do much to reverse the recent hemorrhaging of trust in the all important US-China relationship.

* * * * *

"Beyond Security, towards Peace" was the motto of the Seoul summit. Building upon and further expanding past achievements, the international community had the opportunity to take another step toward a safer, more secure world, free from fear of nuclear disasters either intended or unintended, with South Korea potentially contributing to help formulate a stronger global regime for nuclear nonproliferation.

At Seoul, states had to assess how to incorporate elements of the nuclear-safety regime (e.g., regularized assessments, information sharing, peer review, reviews of the implementation of relevant international conventions, and strong trade organizations) into the nuclear-security regime over time. States had to acknowledge that barriers such as national sovereignty, lack of information transparency, lack of policy consensus, and regime harmonization were significant challenges that needed to be addressed. The international community had to seek an optimal balance between mandatory international standards and voluntary actions and endorse consideration of additional binding and nonbinding international safety and security requirements. There was a need for all nations to strengthen the independence of nuclear regulatory authorities, to harmonize accident/incident reporting parameters and expand information sharing and transparency in a crisis, incorporate security as a fundamental element in new reactor designs, and robustly protect nuclear facilities, including against cyber attack, and to encourage civilian nuclear operators and regulators in engaging with their foreign counterparts on nuclear-security best practices while protecting sensitive information and encourage regularized dialogue and interaction among nuclear operators, regulators, international organizations, and policy experts.[13]

Such moves would make the world a step closer to a nuclear-free world. The nuclear terrorism threat is real and urgent. Securing weapons-grade fissile material is the highest priority action to prevent it. Nonetheless, despite numerous efforts, the current state of global nuclear nonproliferation regime is less than satisfactory. How to make it complete is up to the willingness of the nations that care about the future of humanity. A key component to the success of all of these efforts is the ability, or capacity, of GP nations to implement them. The extension of the GP beyond 2012 is essential. Global threats must be addressed globally. The Cold War legacy that led to the proliferation risks for which the GP was originally created have been a major source of global threat reduction activity since 1992.

---

[13] Kenneth Luongo, Sharon Squassoni, and Joel Wit, "Integrating Nuclear Safety and Security: Policy Recommendations," Center for Strategic and International Studies, Policy Perspectives, December 13, 2011; http://csis.org/files/publication/111214_Integrating_Nuclear_Safety_and_Security_Memo_2.pdf, pp. 5–6.

Chapter 4

Addressing the North Korean Nuclear Threat: A Strategic-Pragmatic-Comprehensive Perspective and Its Policy Alternatives*

Kun Young Park

In January 1994, the director of the US Central Intelligence Agency estimated that North Korea might have produced one or two nuclear weapons.[1] In October that year, the Clinton administration, after a

* This study was supported by the Research Fund, 2012, of the Catholic University of Korea. This study has been published in the *Korean Journal of Global Area Studies* 30, no. 3 (2012).

1 It is not known how many kilograms of plutonium North Korea would require for each nuclear weapon because different designs require different amounts of plutonium. Until January 1994, the US Department of Energy (DOE) estimated that eight kilograms would be needed to make a small nuclear weapon. Thus, the US estimate of twelve kilograms could result in one to two bombs. In January 1994, however, DOE reduced the estimate of the amount of plutonium needed to four kilograms—enough to make up to three bombs if the US estimate is used, and up to six bombs if other estimates are used. On April 22, 1997, US Defense Department spokesman Kenneth Bacon officially stated: "When the US–North Korea nuclear agreement was signed in Geneva in 1994, the US intelligence authorities already believed North Korea had produced plutonium enough for at least one nuclear weapon." This was the first time the United States confirmed North Korea's possession of plutonium: Steven Aftergood and Hans M. Kristensen, "Nuclear Weapons Program-North Korea," Federation of American Scientists (November 16, 2006); http://www.fas.org/nuke/guide/dprk/nuke/index.html, accessed August 23, 2012. The IAEA estimates that eight kilograms are required to produce a nuclear device. Jon Wolfsthal, at the Carnegie Endowment for International Peace, assumed in May 2005 that North Korean weapons, if they indeed existed, would require at least five kilograms of plutonium, yielding a possible North Korean capability of eight to eleven weapons: Wolfsthal, "Estimates of North

series of bilateral negotiations, cut a deal with the North that froze its nuclear-production activities. This agreement remained in place until the new US president, George W. Bush, abrogated it in 2003, accusing the North of bad faith and violating the accord. Since then, the North has conducted two nuclear tests. The first, in 2006, was unimpressive, but the second, in Spring 2009, was more successful. A year later, in April 2010, the United States estimated that North Korea had somewhere between "one and six nuclear weapons."[2] If left unchecked, the United States feared, the North could soon amass a larger and more deadly arsenal.

There is no certainty that North Korea has actually built a working bomb, but the increased amount of fissionable material possessed by the North, along with its recently found uranium-enrichment facilities, has raised security threats around the world. The increased North Korean threat has also raised the fundamental question of why the policy efforts by the international community, including the United States and South Korea, to denuclearize the North have failed.

The purpose of this chapter is to propose what I call a strategic-pragmatic-comprehensive (SPC) approach that I believe can facilitate North Korean denuclearization. In subsequent sections, I discuss the dangers posed by a nuclear North Korea, the poverty of the non-SPC policy approaches that have been pursued by both the Obama and Lee administrations, and policy alternatives based on an SPC perspective.

Korea's Possible Nuclear Stockpile," Carnegie Endowment for International Peace (May 11, 2005); http://www.carnegieendowment.org/2005/05/11/estimates-of-north-korea-s-possible-nuclear-stockpile/6lc, accessed on August 23, 2012.

[2] Hillary Rodham Clinton, Secretary of State, "Remarks on Nuclear Nonproliferation at the University of Louisville as Part of the McConnell Center's Spring Lecture Series," University of Louisville, Louisville, KY, April 9, 2010.

The Dangers of a Nuclear North Korea

A nuclear North Korea, if it is accepted by the international community, will cause a number of latent security problems to surface.[3] First, a nuclear North Korea can deliver a body blow to the international security architecture that has been erected and maintained by the United States over the past several decades. More specifically, a nuclear North Korea would certainly embolden other nuclear aspirants around the world and make the Nonproliferation Treaty (NPT), in effect, obsolete. If these nuclear aspirants, encouraged by the North Korean example, develop and possess nuclear weapons, the United States and other nuclear powers will see their global influence decline, causing world politics to become more unpredictable and unstable.[4] If the United States and others want to keep the NPT in place, with North Korea being allowed to enjoy its nuclear status, they will have to pay enormous amounts to keep other nuclear aspirants from entering the "privileged group."

Second, a nuclear-armed North Korea poses a great security and strategic threat to the United States and its allies in Northeast Asia. The United States has forward-deployed a sizable number of troops in South Korea and Japan, and a nuclear North Korea would make these forces, and, the countries hosting them, vulnerable to

---

[3] Some have suggested that a nuclear North Korea is likely to sell its nuclear material and technology to nations and groups that are hostile to the United States. One cannot rule out this possibility, but, it remains a remote possibility assuming that the North Korea knows what this action means to its survival.

[4] Kenneth Waltz has once suggested that the world will become less dangerous as nuclear weapons spread. However, this is based on his theory of balance of power that has turned out to be untenable after the end of the Cold War. Even in the Waltzian-fancy world, humankind will have to bear with unbearable psychological fear and horror. Kenneth Waltz, "The Spread of Nuclear Weapons: More May Be Better," *The Adelphi Papers*, no. 171 (London: International Institute for Strategic Studies, 1981).

nuclear blackmail. The worst-case scenario is that a threatened and insecure North, believing preemptive attack is imminent, uses its nuclear arsenal against US forces in South Korea and Japan. The Obama administration has generally moved away from the unilateral Bush security doctrine, but, North Korea knows that throughout the Cold War period, Democratic presidents have waged more wars than Republican ones.[5]

Third, a nuclear North Korea could cause a nuclear arms race in Northeast Asia, raising the probability of Sino-Japanese and/or China-Taiwan conflicts. The United States would be most likely to get involved in either of these conflicts, and South Korea, too, because of its alliance commitments to the United States, could be drawn in. The reality is that no nation in Northeast Asia, including North Korea itself, wants to get involved in an unwanted war that may be triggered by a nuclear North Korea. Thus, it is in the interests of all the nations in Northeast Asia to persuade the North to give up its nuclear-weapons program.

Moreover, denuclearization is in the political interest of US President Barack Obama, who has consistently spoken in behalf of a "world without nuclear weapons." It was this lofty vision that helped him win the Nobel Peace Prize in 2010. Obama's resounding international and domestic political support would be seriously undermined by a North Korea with nuclear weapons and the consequences entailed in that.

There are a number of ways to denuclearize North Korea. The

---

[5] President Obama said in an interview that he was carving out an exception for "outliers like Iran and North Korea" that have violated or renounced the main treaty to halt nuclear proliferation. Obama made it clear that nations like North Korea and Iran that have violated or renounced the NPT to halt nuclear proliferation will be excluded from the "Negative Security Assurance" afforded to NPT-compliant nations: David Sanger and Peter Baker, "Obama Limits When US Would Use Nuclear Arms," *New York Times*, April 5, 2010.

use of economic sanctions has been the strategy most often adopted by both South Korea and the United States. In subsequent sections, I review this policy's performance and its consequences. This review sets the context for both my analysis of why sanctions have failed and a discussion of an alternative approach that I believe will facilitate the thawing of the concerned parties' relations and help bring about a solution to the pending nuclear problem faced by all three nations.

Performance of the "Principled" Approaches of the Lee and Obama Administrations

The Lee Approach

In February 2008, a conservative, Lee Myung-bak, was inaugurated as president of South Korea, replacing the liberal Roh Mu-hyun. The Roh government had been at odds with the Bush administration over North Korea policy. Roh preferred a policy of engagement, which he believed would change the North's behavior. Bush profoundly disagreed and pursued a more confrontational policy that, as he phrased it, "would not reward the North for its bad behavior." Although Bush changed course later in his second term, primarily because the US public turned its back on his rigid hardline foreign policy in the November 2006 midterm elections, the fundamental tenet of the hardline stance toward the North nevertheless persisted. The new Lee administration took a stance much more in line with the Bush administration and began to reverse key elements of Roh's approach, including the 10.4 accord that committed the two Koreas to negotiate "a joint fishing area" in the contested waters of the West (Yellow) Sea, and naval confidence-building measures "to avoid accidental clashes and turn it into a zone of peace." The Lee administration also launched, in 2008, a policy called Bihaek-Gaebang-Samchun (No Nuke-Opening Up—3,000 Dollars). This was basically a quid pro quo with the North's first move: *if* the North gave up its nuclear-weapons program and opened itself up to the international community, *then*, South Korea would help it increase its annual per capita income to 3,000 dollars in ten years. While the Lee administration considered its North Korea policy to be reasonable and "principled," North Korea saw it as a betrayal of what it believed the South had promised under Roh. The North defied and disregarded Lee's overtures outright. Many experts on North Korea doubted the new policy's feasibility. As *the Korea Times* editorialized, "Lee set preconditions that the North was unable to accept."[6] The North appeared to have concluded that the Lee government was insincere and had no serious intention of improving inter-Korean relations because its policy was not even close to the principle of "simultaneous actions" that then governed interactions between the North and the United States in multilateral/bilateral negotiations in Beijing.

Inter-Korean relations grew tense and were marked by a series of tragic and provocative events. In July 2008, a North Korean soldier shot and killed a tourist from the South at Mount Geumgang. In April 2009, provocatively, North Korea test-fired a rocket suspected to be a long-range ballistic missile. In May 2009, the North conducted its second underground nuclear test. On March 26, 2010, the South Korean patrol ship *Cheonan* was sunk near the North-South maritime border. Forty-six sailors died. The South ultimately accused the North of torpedoing the ship, although the latter denied involvement in the *Cheonan* sinking. On May 24, 2010, Lee said that South Korea would sever almost all trade with Pyongyang in response to North Korea's sinking of the *Cheonan*. In July 2010, the South initiated a US-

----

[6] Lee Chang-sup, "Inter-Korean Relations 4.0," *Korea Times*, August 11, 2011.

Korea joint military exercise as a demonstration of force. This caused China, North Korea's only ally, to become involved in the dispute. China said that it opposed any US–South Korea military exercises in the entire "Yellow Sea."[7] The Chinese protested so vociferously that the planned maneuvers were shifted to the East Sea (Sea of Japan), east of the peninsula.[8] China was wary of the military drills because it believed that Seoul was trying to involve the United States in these inter-Korean disputes, although the South stated that it would independently make a judgment.

The bad situation soon grew worse. In November 2010, North Korea heavily shelled the South Korean island of Yeonpyeong. Approximately, 200 rounds hit the island, killing 2 soldiers and injuring 17 others, as well as 2 civilians. South Korea returned fire and scrambled combat aircraft in the area. North Korea blamed South Korea for the incident, stating that "the South Korean enemy, despite our repeated warnings, committed reckless military provocations of firing artillery shells into our maritime territory near Yeonpyeong island."[9]

Flabbergasted and angry, the Lee administration attempted to intimidate the North by encouraging the United States to bring the nuclear-powered *USS George Washington* (CVN 73) and its strike group into the US-South Korea joint military exercise area in the West Sea. It was unclear whether the North was intimidated or not, but the North warned of a retaliatory "holy" war against the United States and the South. China expressed great concern about such military activity

---

[7] Foreign Ministry Spokesperson Qin Gang, regular press conference (July 8, 2010); http://www.fmprc.gov.cn/eng/xwfw/s2510/2511/t715219.htm, accessed August 23, 2012.

[8] Jeremy Page, "China Warns US as Korea Tensions Rise: Beijing Bears Rising Cost for Support of Pyongyang," *Wall Street Journal*, November 26, 2010.

[9] KCNA news agency, "North Korean Artillery Hits South Korean Island," BBC News, November 23, 2010; http://www.bbc.co.uk/news/world-asia-pacific-11818005, accessed August 23, 2012.

by "foreign navies" so close to its territory and warned the United States not to venture too close to Chinese territorial waters. Both the United States and China were aware of a possibility of "entrapment" and were finally able to prevent the disputes from escalating. But Lee warned that South Korea would unleash "enormous retaliation" should the North again launch attacks. He also stated that any possible future inter-Korean dialogues would require North Korea's apology for the incursion and its promise not to repeat such provocations.

Thus, since the inauguration of the Lee administration and the adoption of what it calls the "principled approach," inter-Korean relations have swiftly deteriorated to the worst point since the end of the Cold War. However, tensions on the peninsula started to relax as the United States, for strategic reasons, recently changed course. Domestic political considerations may have also played a role in this policy change. With the presidential election looming, the Obama administration wanted to prevent instability on the Korean peninsula from becoming a diplomatic liability.[10] The Lee administration, as it had tried to please the United States under Bush four years ago, followed the US initiative and began to move in the direction of reengaging the North,

---

[10] Iran is the source of this concern, which in turn highlights the danger of the instability on the Korean peninsula that would bring an insurmountable burden to the Obama administration, which has ended the two-war strategy. Meanwhile, the Obama administration suffers from a huge deficit and the fatigue from the wars in Iraq and Afghanistan. And, recently, Obama and the Democrats are aware that they are vulnerable on high gas prices and recognize that the issue could help Republicans sow public doubt about whether the economy is improving. Greg Sargent, "How vulnerable are Dems on rising gas prices?" *Washington Post* (February 24, 2012). Lee is now also responding to domestic political dynamics he had apparently ignored. The Lee administration appears to understand that the preelection political manipulation known as the "North Wind" strategy is likely to backfire as it has done in past years. It is also aware that greater portion of the South Korean people favors improvement in inter-Korean relations. In a survey conducted in early 2012, eight out ten South Koreans(79.8 percent) support inter-Korean stability over confrontation and tension: Hyundai Research Institute, "2012 Survey on Inter-Korean Relations"; http://www.hri.co.kr/board/ReportView.asp.

especially as the United States, in bilateral talks, urged the North to improve its relations with the South.

Those negotiations, however, lost their momentum with the death of "the Dear Leader" Kim Jong-il on December 19, 2011. Furthermore, the imprudent handling of the issue of offering condolences by the Lee administration appeared to have cost South Korea a rare opportunity to reengage with North Korea. What the Lee administration did was relatively forward-looking, depending on how one looks at it, especially when compared with what the Kim Young-sam administration did when Kim Il-sung died in July 1994. The latter refused to offer condolences, but the former expressed its sympathy, although it was direct to the people of North Korea, rather than Kim's family or the North Korea itself. The Lee administration allowed bereaved family members of the late South Korean President Kim Dae-jung and the late Hyundai Group Chairman Chung Mong-hun to visit the North in return for a visit by North Korean delegates to the funerals of the two South Korean figures.

However, North Korea, a nation with Confucian hierarchical culture at its heart and an absolute authoritarian political system centered around the "supreme leader," or "His Majesty,"[11] was enraged at South Korea for refusing to send a government delegation to the funeral. It barred all South Koreans, except for two private delegations, from visiting to pay respects after Kim's death, and, not less importantly, for placing its military on high alert.

It was deja vu all over again. In 1994, when Kim Il-sung died, President Kim Young-sam—who had been due to meet "the Great Leader" imminently in what would have been the first inter-Korean summit, brokered by the US former president, Jimmy Carter—angered

---

[11] On March 4, 2012, thousands of North Koreans rallied to slam the South Korean leader. They criticized Seoul for allegedly slandering "the supreme majesty." "N. Koreans Decry S. Korean Leader at Huge Rally," *Korea Herald* (March 4, 2012).

Pyongyang by putting troops on high alert and sending no condolences.[12] In 1994, North Korea heavily criticized the Kim administration for "acting contrary to human morality."[13] President Kim said he still wanted to hold a summit meeting with Pyongyang despite Kim's death, but the North responded by saying that "it would never deal with such a despicable man." Inter-Korean relations deteriorated sharply and stayed cool for the rest of Kim's term as president.

Regarding the Lee administration's attitude toward the condolences issue, North Korea claimed it to be "an unacceptable insult to and mockery of our dignity" and "a barbaric crime against humanity."[14] It also said that the North Korean army would force South Korea to pay a price for putting its troops on high alert during the national mourning period. The Lee administration, like the previous Kim administration, tried to defuse the situation by stating that South Korea's "basic principle toward easing tensions in inter-Korean relations and resolving problems through dialogues remains still and that it hopes North Korea to regain stability in the near future and present constructive attitude." North Korea's response was simply that "we will

---

[12] When news of Kim Il-sung's death was broken on July 9, 1994, the then South Korean president, Kim Young-sam, offered no condolences to the North Korean leader, saying only at a national security meeting: "I feel regret [over Kim Il-sung's death]." South and North Korea's leaders planned to meet two weeks later to have a frank dialogue on peace on the Korean Peninsula and the future of our nation: "How S. Korea responded to Kim Il Sung's death in '94," *Dong-A Ilbo*, December 21, 2011; http://english.donga.com/srv/service.php3?biid=2011122194908, accessed August 23, 2012.

[13] The Korean Central News Agency was critical, commenting: "At a time when even the heads of the US and Japan express their condolences, only Kim Young-sam, who is the counterpart of the inter-Korean summit and a compatriot, is acting imprudently." Ibid.

[14] North Korea's propaganda website "Uriminjokkiri" commented in an editorial titled "We Are Watching South Korean Government": "N. Korea Blames South for Showing Condolence," Hanguk Story, December 22, 2011; http://hangukstory.blogspot.kr/2011/12/n-korea-blames-south-for-showing.html, accessed August 23, 2012.

have no dealings with the Lee Myung-bak group of traitors forever."[15]

The North's chilly reception notwithstanding, the Lee administration persisted in its effort to change the atmosphere of inter-Korean relations. On January 2, 2011, President Lee, in a nationally televised New Year's address, stated that the Korean peninsula was "at a turning point" and offered North Korea a "window of opportunity" to improve relations, while warning of powerful retaliation if Pyongyang launches another military strike. What was striking in this address was that Lee did not mention either the *Cheonan* ship incident or the shelling of Yeonpyeong Island. In May 2010, he had firmly linked any possible future inter-Korean dialogues to North Korea's apology and promise not to repeat the provocations. His New Year's address was a great contrast.

Moreover, on February 21, 2012, in a press conference marking the fourth anniversary of his inauguration, Lee further toned down his hardline North Korea policy and declared that if North Korea was open to talking "with a sincere attitude," Seoul would "welcome it and talk with an open heart." Interestingly, when Lee was asked by a journalist "if North Korea takes a turnaround approach on its position on the nuclear or human rights issues, is it possible for South Korea can improve relations with North Korea without the latter's apology for the *Cheonan* sinking and the Yeonpyeong shelling?"[16] Lee dodged the question. He seemed be in a dilemma between his need to improve inter-Korean relations and continue to attract anti-North Korean conservatives as the 2012 general/presidential elections. But, actually, since the first day of 2012 it has been apparent that the Lee administration

---

15 The statement was in an English-language version of the KCNA report.

16 Lee Myung Bak, President's Special Address, February 21, 2012; http://www.president.go.kr/kr/president/news/news_view.php?uno=1803&article_no=438&board_no=P01&search_key=&search_value=&search_cate_code=&order_key1=1&order_key2=1&cur_page_no=1&cur_year=2012&cur_month=, accessed August 23, 2012.

has decided to delink the inter-Korean dialogue from North Korea's offering of "remorse." A senior South Korean government official has already said that an admission from North Korea that it sank the *Cheonan* or an apology for its shelling of Yeonpyeong Island are "not a precondition to the resumption of the six-party nuclear disarmament talks."[17]

The Lee administration's dropping of the apology requirement appeared to have led to nuclear negotiations between the North and the South, raising cautious hopes that after months of recriminations the countries were inching toward broader talks on ending the North's nuclear-weapons program. The nuclear negotiators for South and North Korea had met in 2011 in Bali and Beijing, suggesting that the two sides were moving to a new stage of dialogue. But they were not able to produce any tangible agreement that could move the multilateral nuclear talks forward.

The Obama Approach

The hardline North Korea policy pursued by the "neoconservatively principled" Bush administration, which helped abolish the Agreed Framework in 1994, faced a problem in early 2005 when the North declared that it had nuclear weapons. That declaration helped to create the dynamic by which the North engaged in direct negotiations with the Bush administration. These talks produced a couple of significant accords, such as the 9.19 and 2.13 (the action plan of the 9.19) agreements. But, the last round of six-party talks ended with a stalemate on December 11, 2008 because North Korea refused to agree in writing that verification of its nuclear activities would include scientific sam-

---

[17] "Apology No Precondition for Talks with N. Korea, Seoul Says," *Chosun Ilbo*, March 11, 2012.

pling. The North insisted that "it is only required at this point to carry out the limited verification steps agreed in writing with the United States in October 2008, which did not include sampling provisions."[18] The North apparently decided to wait until a new US president took office.

However, the new US president was bombarded with pressing domestic/international problems that required his immediate attention as he was inaugurated in January 2009. The Obama administration gave a priority to such domestic issues as economic recovery and revamping of US health care systems. Its foreign policy focus was on those issues that were more of a direct concern to its domestic political interest—curbing global terrorism, mopping up Al Qaeda in Afghanistan/Pakistan, nation-building in Iraq, and managing the Iran-Israel conflict in the Middle East. Impatient, and perhaps frustrated at being placed on the back burner, North Korea appeared to "test the waters" with the United States by launching a long-range rocket in defiance of international pressure on April 5, 2009. Obama responded by bringing the matter before the UN Security Council, which subsequently tightened its sanctions on the North. Apparently, this US move led the North Koreans to think that the new administration would not pay them the serious attention they thought they deserved. Out of frustration, and myopic impetuousness, North Korea went ahead and tested a second nuclear device in May 2009, expelled US and international nuclear inspectors, and declared it would "never" return to the "six" talks.

The Obama administration did not handle this obnoxious development effectively. It took several months to come up with a policy on these North Korean provocations. In December 2009, the

---

[18] Peter Crail, "Six-Party Talks Stall over Sampling," *Arms Control Today* 39, no. 1 (January/February 2009).

administration finally articulated what Secretary of State Hillary Clinton dubbed a "strategic patience" policy. This essentially was to wait for the North to come back to the talks while sustaining pressure by way of economic sanctions and arms interdictions.[19] On July 21, 2010, the United States imposed new sanctions against Pyongyang for its involvement in the sinking of the *Cheonan*, and from July 25 to July 28, 2010, the United States and South Korea conducted a joint military exercise in the East Sea as a show of force. On August 30, 2010, President Obama signed an executive order increasing financial restrictions on North Korea.

It was not always like this. North Korea came "tantalizingly close" to a deal with President Bill Clinton in his final days in office in 2001.[20] According to a former senior US official, the agreement, when completed, would have both halted North Korea's exports of missiles and related technology and stopped further production, deployment, and testing of long-range missiles. New President George W. Bush refused to strike a deal with the North and had Secretary of State Colin Powell retract his statement that "the Bush administration would pick up where Bill Clinton left off in negotiations with North Korea." It is ironic that, in 2009, a Democratic president, who preached to the world the importance of diplomacy, was unable to harvest a crop cultivated by a Republican president who once declared that "It's better to be feared than respected." In any case, the United States and North Korea had two opportunities, one in 2001 and one in 2009, but made grave blunders and lost chances to help defuse their respective security concerns.

In 2010, North Korea apparently felt the economic pressures

---

19 David Gollust, State Department, "Clinton Calls 'Exploratory' Meeting with North Korea 'Quite Positive'," *Voice of America*, December 10, 2009.

20 Wendy R. Sherman, "Talking to the North Koreans," *New York Times*, March 7, 2001.

brought on by the Obama administration's strategic patience policy, but it hardly showed signs of serious stress. It was rather the United States that changed course due to the emergence of an unexpected development, as it had in February 2005 under Bush. Again, the United States had to react to a North Korean initiative. In November 2010, the North invited a team of US nuclear scientists and North Korea specialists to North Korea to reveal to them the 2,000-centrifuge uranium-enrichment facility it had constructed. It claimed that the facility would produce Low enriched uranium (LEU) for a light-water reactor (LWR), which it also revealed was under construction. Pyongyang admitted for the first time that it could produce uranium hexafluoride (UF6), the feedstock for uranium enrichment, confirming long-held suspicions about it having such a capability. On February 16, 2011, in Senate testimony, the US director of national intelligence, James Clapper, said that North Korea likely had additional undeclared uranium-enrichment facilities beyond what it had first revealed in November 2010.

This new development put "diplomacy" back to work. Following the first round of denuclearization talks between the nuclear negotiators of the two Koreas, the United States invited the North Korean vice foreign minister, Kim Gye-gwan, to New York in July 2011. The New York bilateral talks, the first since six-party talks collapsed in 2009, dealt with the issue of resuming negotiations over the North's nuclear program in exchange for food aid by the United States These led to further negotiations in Geneva that the US representative described as "moving in the right direction." Although there were differences that needed to be narrowed further, the news media in North Korea reported that Kim Jong-il made rare comments on the possibility of resuming six-nation talks.[21]

---

[21] Brian Knowlton and Choe Sang-hun, "New US Envoy to Talk with North Koreans," *New York Times*, October 19, 2011.

But in April 2012, the new North Korean leader, Kim Jong-un, went ahead and tested a long-range rocket, defying the international community's request for restraint, making the 2012 North Korea–US accord obsolete. Hardliners reemerged in both South Korea and the United States requesting a rapid arms build-up, including redeployment of tactical nuclear weapons in the South. The North's reason for choosing what the international community considers a confrontational approach is not entirely clear, but an anti–North Korea newspaper in the South indicates a possibility of miscommunication between the North and the United States. According to the *Chonsun Ilbo*, North Korea claims it informed the United States of plans to launch a space rocket in July 2011. The newspaper suggests that in the letter it obtained right after the North's announcement of the imminent launch of the rocket, North Korea's Kim Gye-gwan claims that Pyongyang mentioned its plan to propel a satellite into orbit during the first high-level meeting between the United States and North Korea. Specifically, Kim wrote: "Since the first bilateral, high-level meeting, I made it clear that the long-range missile moratorium does not include satellite launches for peaceful purposes. And this provided a foundation for agreement between North Korea and the US." Kim expressed concern that the United States could halt food aid, which would be "a clear violation" of the essence of what the two countries agreed. "Our relevant institute has already started technical procedures to stop the operation of the centrifuges at the uranium-enrichment facility in Yongbyon," he added, indicating that the North was abiding by the bilateral agreement. Kim then proposed early further talks in Beijing or other convenient place to "personally" explain the North's stance and discuss ways to control the situation following the rocket launch.[22] The information in the

---

[22] "N. Korea 'Told US About Rocket Launch Last Year,'" *Chosun Ilbo*, April 12, 2012; http://english.chosun.com/site/data/html_dir/2012/04/12/2012041201340.html, accessed August 23, 2012.

Chosun Ilbo seems reliable precisely because it has maintained a staunch anti–North Korea stance since the partition of the Korean peninsula. Therefore, assuming that the United States is not lying, there must have been some kind of a communication failure between the two nations. It is probable that the United States may have said that it cannot accept the North's launch, but the latter may have assumed that its position was understood by the United States, which apparently did not seriously challenge it.

The Poverty of Lee's and Obama's Policies on North Korea

This brief historical review of inter-Korean and US–North Korea relations has shown that not only have their policies toward North Korea not been productive, but that they are close to a failure. They have not prevented North Korea from increasing its nuclear capability and they have undermined peace and stability on the peninsula. While the Lee administration is now trying to find a way to improve inter-Korean relations, through what the unification minister has called a "flexible approach," it is apparent that it is not likely to get what it wants, given the aggressive and recalcitrant postures of the North. Furthermore, it seems virtually impossible for the administration to take the driver's seat in dealing with North Korea, which was a national foreign-policy objective for previous administrations. The Obama administration, on the other hand, was most recently able to rekindle the embers of a dialogue. However, concluding a viable working agreement with the North is still a dream, especially as North Korea's position on the nuclear issue seems to have become more ambitious and swaggering.

Of course, North Korea is far more responsible than either South Korea or the United States for the deterioration of their relations and for the dangers they have caused. However, the North's noncooperation or "defection" cannot justify the North Korea policy of the Lee and Obama administrations. The North's "bad" intentions and behaviors have persisted over the years since the partition of the peninsula, and not just over the Lee and Obama years. What is noteworthy, however, is that some previous South Korean and US administrations were arguably able to constrain the North's "bad" behaviors and create, and benefit from, stable and cooperative inter-Korean and US–North Korea relations. The most significant and proper words that characterized the inter-Korean relations during the earlier two Korean administrations were the inter-Korean summit meetings, the Gaesung Industrial Complex, and Mount Geumgang tourism, rather than the sinking of a patrol ship and the shelling of Yeonpyeong island. They were, in the case of US-North relations, the October 2000 visit to the White House by Vice-Marshal Cho Myung-rok and a reciprocal visit to Pyongyang by US Secretary of State Madeline Albright, rather than the long-range ballistic-missile and nuclear tests that insulted the United States It tells a lot about how the North Korea policy of South Korea and the United States can make a difference.

That said, one can readily understand how the question of "who is more to blame" can be a useful tool for some politicians looking to score domestic political points. They can claim that they wanted to improve the relations based on a principle of reciprocity, which they argue is a norm in international relations. But, despite their good efforts and intentions, the North rejected their offers. They claim they have stood on their principles, and, therefore, they deserve the people's support. And this political strategy works. It is undeniable that this creates domestic political benefits. But to make North Korea (or any foreign) policy serve someone's domestic political purpose is both noxious and absurd. It sacrifices national interest for the benefit of domestic political capital.

That question is also unwise if one looks at it from the perspective that the South needs to actively create, initiate, and lead a peace process on the peninsula that would ultimately bring about peaceful and democratic unification. In this view, what is crystal-clear is that the Lee administration has failed to recognize the role it must to play and, consequently, has failed to manage inter-Korean relations in a way that will benefit both parties. For his part, Obama has been too passive as an ally of South Korea, whose interest is in peace and stability on the peninsula. He was unable to meet the expectations he raised as a visionary pragmatist during his 2008 presidential-campaign period.

Why the Policies Failed

The North Korea policies of both the Lee and Obama administrations have failed. And the reasons for these failures are, first, that there was a lack of prudent conceptual framework with regard to North Korea policy. As a result, neither administration had clear policy objectives. One example of this is Lee's early attempt to abolish the unification ministry that had operated with the assumption that inter-Korean relations had special characteristics that derive from the fact that North Korea is both the South's "enemy" and "friend" at the same time and that it is also the target of the South's unification effort. The significance of this attempt, based on a recommendation of the presidential-transition committee,[23] to marginalize the ministry as a factor in policymaking under the new administration, was not lost on the

---

[23] Lee's transition team wrote in a report unveiling the proposal that "the issue of unification should be handled consistently with foreign policies" Burt Herman, "S. Korea to scrap N. Korea-focused Ministry," *USA Today*, January 16, 2008.

North. It became the immediate pretext for the North to publicly amplify its criticisms of Lee and the new administration.[24] The unification ministry was ultimately saved due to social and political pressures, but what it did was almost nothing. It played the role of a Cold War warrior-style ideologue that promoted the idea of confrontation and containment, rather than peaceful coexistence and cooperation with the North. Unlike Lee, Obama did not have a reason to regard the North as "something special." But it seems clear that he did not understand that global/regional peace and stability derives as much from the prudent management of North Korea as it does from the proper handling of conflicts in the Middle East.

The ideological rigidity is the second item revealing the poverty of Lee's and Obama's North Korea policies. They both declared, directly or indirectly, that they would be pragmatic in designing and implementing their foreign policies. In reality, however, both pursued policies designed to pressure North Korea into changing its system and to open up the country to the capitalist and liberal outside world. South Korea and the United States knew how to criticize North Korea for its wickedness, but they did not know how to "beat it at its own game," pragmatically and strategically taking advantage of its wickedness and capitalizing on it.

Third, their fixation on the utility of sanctions was quixotic, to say the least. For example, Obama's policy of strategic patience, which Lee actively supported, focused on economic sanctions and the diplomatic isolation of the North. This option turned out to be innocuous and almost senseless when applied to North Korea. The United States and its allies have long imposed severe sanctions on the North, with little result. They can hardly bring "more" pressure on it. The effect

---

[24] Scott Snyder, "Lee Myung-bak's Foreign Policy: A 250-Day Assessment," *Korean Journal of Defense Analysis* 21, no. 1 (2009).

would be like denying meat to a Buddhist monk. Rachel L. Loeffler, formerly at the US Department of the Treasury, has suggested that "financial sanctions taken against Iran and North Korea make clear that this new financial statecraft can be effective."[25] What she says is basically that the "squeeze on Banco Delta Asia hit North Korea where it hurt."[26] There may be some truth to this, but it is doubtful whether such financial sanctions can force the North to change. In fact, the US financial sanctions in 2007 ended in compromise because the United States chose to arrange the transfer of the North's funds on deposit at Banco Delta Asia to a bank in Russia from which the North could withdraw its money. It is important to remember that North Korea can learn from history. It therefore remains unclear whether the financial sanctions can repeat their effect in the future. It also seems improbable that sanctions would instigate popular opposition to the ruler in North Korea: the most potent forces of radical political change in developing nations have been college students and the military. But in the North, these actors are themselves the privileged elites committed to preserving the existing political/social order and their place in it. Moreover, the option of sanctions backfired. The pressure has increased the North Korean people's suffering and strengthened the pretext that the North Korean ruler would use to solidify his principle of "Military-First Politics." If unintended effect accumulate, North Korea will be more dangerous and miserable than the international community can ever handle.

Probably the most important factor that has neutralized the option of sanctions is related to the lack of Chinese cooperation. For example, the UN sanctions on the North after its 2006 nuclear test

---

[25] Rachel L. Loeffler, "Bank Shots: How the Financial System Can Isolate Rogues," *Foreign Affairs* 88, no. 2 (March/April 2009).

[26] David Lague and Donald Greenleesm "Squeeze on Banco Delta Asia Hit North Korea Where It Hurt," *International Herald Tribune*, January 18, 2007.

has had no perceptible effect on North Korea's trade with China.[27] The Chinese attitude showed no change even after the North's second nuclear test in 2009. In fact, this nuclear test dramatically raised the stakes in the global effort to persuade a recalcitrant North Korea to give up its nuclear-weapons program, and even China joined in the criticism with a strong statement expressing its "resolute opposition" and demanding that "North Korea keep its promise of denuclearization and cease all actions that could further worsen the situation."[28] China watchers in the West wondered if Chinese anger over a nuclear test might lead it to change its policy toward the North.[29] As a matter of fact, the North Korean provocation seemed to prompt China to reconsider its North Korea policy. The Chinese authorities unprecedentedly allowed heated debates among the Chinese intelligentsia about how to handle North Korea.[30] The *Global Times* (*Huanqiu Shibao*), a sister newspaper of the *Renmin Ribao*, published a survey of twenty Chinese experts on international relations and found them evenly divided between those that supported more severe sanctions on North Korea("the strategists") and those that opposed such actions("the traditionalists").[31] But the Chinese politburo disregarded the strategists' argument and sent Premier Wen Jiabao to Pyongyang to sign a series of agreements that promised aid for the North and an expansion of economic exchange.

---

[27] Marcus Noland, "The (Non-) Impact of UN Sanctions on North Korea," *Asia Policy*, no. 7, January 2009.

[28] Press Conference by Chinese MFA spokesperson, May 25, 2009; www.fmprc.gov.cn.

[29] Jay Solomon, Ian Johnson, and Gordon Fairclough, "China's Anger at North Korea Test Signals Shift," *Wall Street Journal*, May 29, 2009; Adam P. Liff, "US Policy toward North Korea: The China Fallacy," Pacific Forum of the Centre for Strategic and International Studies, PacNet #67, no. 8 (October 2009).

[30] In 2004, the PLA-related journal *Strategy and Management* was shut down because it published an article criticizing North Korea's leadership, evoking fierce protest from Pyongyang.

[31] Tang Xiang, "Many Noted Chinese Scholars Support More Severe Sanctions to Be Imposed on DPRK," *Huanqiu Shibao Online*, May 26, 2009.

The Chinese understood that Wen's trip would be intensely monitored by the rest of the world, looking for indications that China would pressure the North to halt its nuclear-weapons program. That is what makes the agreements all the more significant. The deal struck by the two nations clearly reveals their shared primary goal—stability in North Korea. It is true that the internal debates have prepared the ground for possible future changes in China's North Korea policy, but in the short run the current policy is likely to persist as long as the traditionalist argument remains in sync with the international and strategic reality perceived by the Chinese government.[32] Sanctions against North Korea are not likely to bring about the desired results hoped for by South Korea and the United States. Chinese cooperation is essential, and that will hardly be forthcoming in the near future.

In sum, the hardline or "principled" North Korea policies of the Lee and the Obama administrations have failed. The lack of a clear conceptual framework, ideological rigidity, "domestic politics pandering," strategic miscalculations, and wishful thinking have all combined to produce a situation in which the two Koreas bitterly face each other and the North is acquiring more information on nuclear weaponization, making more nuclear weapons, and churning out more dangerous fissile materials. This analysis of the poverty of the North Korea policies of South Korea and the United States has been brief, but it serves to set the stage for a discussion of the policy alternatives available to the two nations.

---

[32] The International Crisis Group judiciously summarizes China's largest concerns as the possibility of military confrontation between North Korea and the United States as regime implosion, a flood of North Korean refugees into China, or precipitous reunification with South Korea leading to a US military presence along the border: "Shades of Red: China's Debate over North Korea": *Asia Report* no. 179, November 2, 2009; http://www.crisisgroup.org/en/regions/asia/north-east-asia/china/179-shades-of-red-chinas-debate-over-north-korea.aspx, accessed August 23, 2012. From a broader, strategic perspective, the encirclement strategy China believes the United Sstates. has been pursuing constitutes a serious threat to the security of China. In sum, the North Korea policy of China heavily depends on the Chinese perception of the China policy of the United States.

An Alternative Approach Based on the Strategic-Pragmatic-Comprehensive Perspective

Denuclearization, along with peace and stability in the Korean peninsula, has been a core objective of North Korea policies of South Korea and the United States But one is faced with a fundamental question before proceeding to think about how to achieve such goals: "Will North Korea ever give up its nuclear-weapons program?" This is difficult to answer because it requires access to the thinking of the North Korean leadership, which outsiders do not have. But it is possible to make an educated guess if one is familiar with the historical patterns of its foreign policy decision making and the political, strategic, and economic situations that help shape the North's decisions. My educated guess is that, in the short run, it is hardly possible for the North to change course on nuclear weapons—primarily because no North Korean leader is free, or wants to be free, of the "glorious" legacy of Kim Jong-il, "the sun of the nation." Political legitimacy and, therefore, political power, comes from him. The more the current leaders need power, the more they need to be identified as faithful successors of Kim Jong-il. This unique transitional politics is reflected in the editorial carried in the North's three major state-run publications early this year: "It is the steadfast determination of our Party that it will make no slightest vacillation and concession in implementing the instructions and policies [Kim Jong-il] had laid out in his lifetime." In the same vein, North Korea has proclaimed itself as a nuclear power in its new constitution, which was amended April 13 by the Supreme People's Assembly.

The most significant of Kim's policies was to build a "prosperous and strong nation," putatively by April 15, 2012, the 100-year anniversary of the birth of Kim Il-sung. The most significant political mechanism undergirding that notion is the Military-First Politics, the objective of which is to make North Korea a nuclear-armed nation. Since his rise to the top leadership position in the North, Kim Jong-un has been depicted in the North's state-run media as the faithful successor of his father's Military-First Politics. As a symbolic gesture of buttressing that policy, Kim Jong-un, in his first public action as supreme commander, inspected "the Seoul Ryu Kyong-su 105 guards tank division," the first North Korean unit to enter Seoul during the Korean War. This is an insinuation that the North Korean leadership cannot and will not give up its nuclear-weapons program until the long shadow of Kim Jong-il fades.

This point begs another question regarding the future of the North's nuclear-weapons program. Is it possible for the North's leadership to change course if/when it finds the "balance of needs" shifting from the political to such economic and social needs as food and energy? I would say that it is possible, although it will be more difficult for the international community to have the North change course than it would have been in the past. One key reason why it has become more difficult to influence change in the North is that it now believes that its two nuclear tests have advanced it to the rank of a nuclear power. On April 21, 2010, North Korean state media reported that Pyongyang issued a memorandum stating that the country would be party to nonproliferation and disarmament agreements "on an equal footing with other nuclear weapons states." The North has put a much higher price tag on its nuclear-weapons program.

A second reason is that the experience of both Iraq and Libya will make North Korea extremely cautious in making decisions on denuclearization. As one North Korean official phrased it, "it is now being fully exposed before the world that Libya's 'nuclear dismantlement,' much touted by the United States in the past, turned out to be a mode of aggression, a way of coaxing the victim with sweet words to disarm itself and then to swallow it up by force."[33] Denuclearization will be possible only if and when the North is confident that it can be safe without nuclear weapons.

Although the solutions to the nuclear problems have become far more difficult to find, they are still possible if the international community can put together the right combination of prudent policies and have the consistency to follow through on them. The United States and South Korea are the most relevant actors on the nuclear issue, and they need to cooperate and coordinate with each other in terms of sharing the objectives, values, vision, burden, and benefit to effectively and efficiently encourage the North to change. Below I discuss a perspective that I think the two nations should share to that effect.

The Strategic-Pragmatic-Comprehensive Perspective

It is important to understand that there must be a "paradigmatic" shift in the thinking of South Korean and US policymakers on North Korea. Then a set of carefully designed methodological adjustments should be made in accordance with the paradigmatic change. I suggest that this new perspective should be pragmatic, strategic, and comprehensive in nature. Each of these is notions delineated below.

Being pragmatic, as I see it, involves an objective rather than subjective (or realistic rather than idealistic) attitude toward the target of negotiations. In other words, as historian E. H. Carr suggested many years ago, a "pragmatic" South Korea and the United States must look at North Korea "as it is" rather than "as it ought to be." In turn, looking at someone "as he is" involves "putting yourself in his shoes (易地思之)." Hans Morgenthau, the great international relations

---

[33] Selig S. Harrison, "Nuclear Weapons and the Libyan Intervention," *National Interest*, April 11, 2011.

theorist, once cautioned that "diplomacy must look at the political scene from the point of view of other nations." He went on to quote the eminent 18th-century British parliamentarian Edmund Burke: "Nothing is so fatal to a nation as an extreme of self-partiality, and the total want of consideration of what others will naturally hope or fear."[34] I stress that this is the core of pragmatic diplomacy, and it is worthy of the attention of South Korean and US decisionmakers who want to change North Korea.

Pragmatism also involves a weighing of benefits and costs and a calculation of probabilities with a particular emphasis on consequences. To cite Morgenthau again, he stressed that successful diplomacy should be divested of any "crusading spirit." Policymakers must keep the objective of foreign policy defined in terms of the "substance of real advantage." My perspective is more than a "value-free" calculus of immediate costs and benefits, however. It considers democratic principles and humanistic values to be the fundamental base on which every foreign policy, or every public policy, for that matter, should be designed and pursued. At the same time, it appreciates the importance of prioritization of issues, and distinguishes between what can be accomplished now and what can be achieved in the future. It highlights the wisdom that problems that seem intractable now will become far easier to solve thanks to the accumulation of prior accomplishments. Thus, my perspective is closely linked to principles and values insofar as the collected pragmatic achievements will work together to contribute, in the longer term, to the realization of such principles and values. It is in this sense that my perspective is strategic as well as pragmatic.

My perspective emphasizes the importance of being compre-

---

[34] Hans Morgenthau, *Politics Among Nations*, 5th ed. (New York: Alfred A. Knopf, 1973), pp. 542–543.

hensive in dealing with the North Korea on nuclear issues. In other words, I suggest that the feud over the North's nuclear-weapons program is not about the weapons program itself—because North Korea believes it is a the matter of national survival. North Korea is determined to keep its nuclear-weapons program because it believes that it would be defenseless before its enemies without it. Therefore, to be successful in denuclearizing the North, I would argue, South Korea and the United States must be attentive to those factors that the North believes most threaten its security. This aspect of my perspective is closely related to the "objective attitude" mentioned above. It is imperative that, in Morgenthau's words, the two nations look at the political scene from the point of view of North Korea. Only then will their "peaceful" North Korea policies bear fruit.

Policy Alternatives

As indicated earlier, the April 2012 rocket launch by North Korea put a damper on the conciliatory atmosphere developing between the North and the United States It will surely take time before "the rocket launch dust" settles down. But the cooling-down period may be shorter than one can anticipate because the 2012 launch hassle seems to have derived from a communication failure between the North and the United States, domestic political considerations, and/or bureaucratic bickering in North Korea, all of which appear less hostile than the reasons behind previous missile launches.[35] As early as July 2012,

---

[35] It is highly probable that the new North Korean leadership thought that the consolidation of the new regime was more important than aid from the United States, and also more important than improved relations. In other words, the launch was more political propaganda bravado, and, more importantly, to placate an unhappy North Korean public at a time when the strong and prosperous great nation was being promulgated. But the leader did not have much to show to his people except launching a rocket. As for the internal bickering, Christopher Hill states that

there were indications that the North was interested in reengaging with the United States in negotiations. These signs included an unusual performance by the Moranbong band for Kim Jong-un that featured Disney characters such as Mickey Mouse and Winnie the Pooh. The inclusion of characters from the United States, the North's wartime enemy, is a notable change in direction for performance arts in the North, and the timing of the performance seems to carry a subtle but important meaning. In any case, this was not the first launch, and there is no viable alternative to a negotiated solution. Therefore, a reengagement between the two nations over nuclear issues appears inevitable, although it is uncertain which side would first hold out its hand. Anticipating the time when the two parties are ready to talk, I suggest that, in order to further induce the North to return to the six-party talks and cooperate with the United States and South Korea in denuclearizing the peninsula, the two allies need to assemble a set of carefully-thought-out policy alternatives that are acceptable and palatable to all of the negotiating parties.

The Lifting of Sanctions

Under the assumption that preliminary talks could bring North Korea and others back into multilateral talks in Beijing, I suggest that other confidence-building steps by South Korea and the United States would be helpful in moving that process forward. One of these could be the lifting of sanctions imposed on the North. It does not make a lot of sense (especially to the North) if, on the one hand, the United States considers the North to be a legitimate negotiating partner, while, on the other, treating it as a criminal that needs to be severely punished.

"people doing the negotiating on food aid are not the same as those in charge of missile launches." Chico Harlan and William Wan, "US and Others Condemn North Korea's Planned Rocket Launch," *Washington Post*, March 16, 2012.

When President George W. Bush called the North "evil," it made it politically difficult to engage the North in dialogue and negotiation even when he saw the engagement to be beneficial to him. "Evil" cannot be any kind of partner for the "righteous" and "moral." And it made compromise almost impossible. I suggest that if the Obama administration finds it politically inconvenient for it to lift the sanctions in full, at least a partial (or gradual) reduction would serve the purpose of creating a minimal level of mutual confidence.

South Korea Should Reach Out to North Korea

Another step would be for South Korea to reach out to the North to improve inter-Korean relations. Again, it would be foolish to argue over who is more responsible for the worsened relations. The "blame game" is one nobody can win. As an architect of peace and stability in the peninsula, South Korea should initiate and lead. South Korea should not forget the missed opportunity to reengage the North when it refused to send official condolences to the authorities in Pyongyang. One can think of at least two reasons behind such a decision. First, the Lee administration might have been sensitive to a possible domestic political backlash, as happened in South Korea when Kim Il-sung died in 1994. But this seems groundless because the political conditions have changed greatly in the past eighteen years. According to a nationwide survey conducted immediately after the death of Kim by the *JoongAng Ilbo*, almost two-thirds (65.4 percent) of South Koreans said the government should express condolences to the North Korean government for the death of Kim Jong-il. Only one-third (34 percent) opposed. These results are a marked contrast to the fierce public debate that erupted in the South following the death of Kim Il-sung in 1994, when most opposed offering condolences or sending a delegation to the funeral.[36]

Second, therefore, it is more probable that the Lee administration did not really want to show "respect" over the death of the "ruthless dictator." It might even have thought that showing respect would be unjust and immoral. It might be understandable on a personal, ethical level, but Lee should have realized that sending condolences had less to do with his personal ethics and much more to do with his political and diplomatic leadership. Morgenthau, again, offers specific advice with regard to exactly such a situation:

> Universal moral principles cannot be applied to the actions of states in their abstract universal formulation. They must be filtered through the concrete circumstances of time and place. The individual may say for himself: "Fiat *justitia, pereat mundus* (Let justice be done, even if the world perish)," but the state has no right to say so in the name of those who are in its care. While an individual has a moral right to sacrifice himself in defense of a moral principle, a state has no right to let its morality get in the way of successful political action.[37]

When Chaiman Mao Zedong died in 1976, US President Gerald Ford and his predecessor, Richard Nixon, praised Mao, despite his significant contribution to US casualties in the Korean War. The reason was that Nixon and Ford wanted the newly opened door to China to stay open. We now know that these "anti-Communist" presidents' decisions served the US interest in that they not only brought a huge economic benefit to the US business community but also helped change a "veiled and dangerous" China into a more open and reformed China. When Kim Il-sung died in 1994, President Clinton issued a statement extending "sincere condolences to the people of North Korea" and

---

36 "Unlike '94, public supports offering condolences," *Korea JoongAng Daily*, December 23, 2011; http://koreajoongangdaily.joinsmsn.com/news/article/article.aspx?aid=2946030, accessed August 23, 2012.

37 Hans Morgenthau, *Politics Among Nations*, p. 12.

expressing appreciation for Kim's leadership in resuming talks with the United States. It was clearly in the US interest to thank him, for it encouraged Kim's successors to continue along the path of adaptation and accommodation. Bob Dole, who was preparing to run for the Republican presidential nomination in 1996, complained that Clinton should have reminded the world of Kim Il-sung's "brutal dictatorship" and his responsibility for a war that cost "54,000 American lives and 100,000 Americans wounded." Shortly afterward, the *New York Times*, in an editorial, said to Dole: "It's called Diplomacy, Senator."[38]

President Lee missed such an opportunity in December 2011. But other opportunities will come along, as long as the United States considers the improvement in inter-Korean relations to be a condition for advancement in US–North Korea relations. South Korea must be ready for this. Park Geun-hye, one of the Korean presidential front-runners, struck a chord with Koreans when she said

> The common strands of the July 4th, 1972 Joint Communiqué, the 1992 South-North Basic Agreement, and the Joint Declarations of June 15th, 2000 and October 4th, 2007 are the principle of mutual recognition and to work jointly for peace. Seen from these perspectives, the Saenuri Party and I firmly believe that both South and North Korea should follow this spirit of "mutual respect and recognition."[39]

North Korea has been distancing itself from the Basic Agreement while emphasizing the significance of the 6.15 and 10.4 accords. The Lee administration has been reluctant to respect the two agreements signed by its liberal predecessors. Although the North Korea policy blueprint that Park published in *Foreign Affairs* in 2011 is full of

---

[38] "It's Called Diplomacy, Senator," Editorial, *New York Times*, July 12, 1994.

[39] Park Geun-hye, chairperson of the Saenuri Party, "Trust-Building Process and the Making of a New Korea," A Keynote Speech, Korean Association of International Studies, February 28, 2012.

rhetoric and wishful thinking and lacks concrete methods for drawing commitments from the North,[40] Park's stance that "all" inter-Korean agreements should be respected also seems balanced, fair, and "paradoxically realistic." Moon Jae-in, another Korean presidential front-runner, who served as chief-of-staff for President Roh, who signed the 10.4 accord, is believed to support all the inter-Korean agreements. Other major presidential runners seem to have taken a similar stance.

One opportunity for engagement that is likely to appear in the near future involves the so-called peninsula pipeline. This project that has been put on the table by Russia's state-owned gas corporation, Gazprom, which hopes to build a pipeline linking Russia's Sakhalin Island to South Korea.[41] For the first time in decades, and despite the recent deterioration of inter-Korean relations after the death of Kim Jong-il, Moscow, Seoul, and Pyongyang all have a strong common interest. If completed, the pipeline would be capable of delivering 10 billion m^3 of Russian natural gas per year to South Korea, which imports 97 percent of its natural-gas needs. A steady supply of natural gas from Russia is increasingly appealing to South Korea at a time when revolutions across the Arab world and the sanctions on Iran are challenging the stability of traditional hydrocarbon suppliers. But since the pipeline crosses North Korean territory, the South is worried that it will give the North an additional tool for political blackmail. In the event of a crisis with Pyongyang, South Korea is asking Gazprom to commit to deliver liquefied natural gas (LNG) by ship from Sakhalin Island.

North Korea, too, would benefit greatly from this project. Its

---

[40] "Can Park Geun-hye's N. Korea Policy Work?" editorial, *Chosun Ilbo*, March 11, 2012.

[41] China was negotiating with Russia over a different pipeline linking Siberia to China through the Russian republic of Altai. But China has put the project on hold, asking for a better price: Sebastian Falletti, "Peninsula Pipeline: Gas Supplies from the Russian Far East," *Jane's Intelligence Review* (January 2012).

economy shrank for the second year in a row in 2010, and it is in urgent need of additional revenue for not only economic but also political reasons. If realized, the Gazprom pipeline project could provide the North with between US$100 and US$200 million of transit fees annually. For the South, this project, with its potential of not only bringing tangible economic benefits, would also offer the Seoul a chance for political rapprochement with the new North Korean leadership. The anticipated diplomatic progress between North Korea and the United States toward the resumption of the six-party talks could pave the way for a gas deal between the two Koreas. When such an opportunity arises, the South Korean leadership should grab it.

Scientific Sampling in the Third Stage

Once the concerned parties return to negotiation, they are expected to pick up where they left off. The six-party talks ended on December 11, 2008, in stalemate because the United States believed that the new agreement with the North would allow inspectors to carry out "scientific procedures" such as sampling, whereas the North denied that it agreed to allow inspectors to carry out sampling at its nuclear facilities.[42] Unlike the Bush administration, the Obama administration appears more flexible over the issue of when the sampling is to take place. Considering that the sampling may instantly reveal what the North might have wanted to conceal before getting what it wants from the United States, it is likely to accept that the agreement on sampling will be negotiated in the third or later phase of the denuclearization

---

[42] On November 13, 2008, a North Korean Foreign Ministry statement said inspection activities are limited to "field visits, confirmation of documents, and interviews with technicians": "Chronology of US-North Korean Nuclear and Missile Diplomacy," Arms Control Association (April 2012); http://www.armscontrol.org/factsheets/dprkchron, accessed August 23, 2012.

road map. In fact, despite the lack of agreement on sampling during the last talks, the then secretary of state, Condoleeza Rice, stated in December 19, 2008, that "within the context of the six-party talks, you ultimately will get a verification protocol that allows us to deal with a lot of very troubling activities." North Korea appears to have left the door open to a written agreement once the October 2007 agreement has been fully implemented. A November 14, 2008 statement by a North Korean Foreign Ministry official suggested that Pyongyang might be willing to address the question of sampling in the third and ostensibly final phase of the negotiations.[43]

Carry Through the 9.19 and 2.13 Accords

Beyond the issue of a verification protocol, the participants in the six-party talks will be engaged in discussions in accordance with the reactivated 9.19 and 2.13 accords, which were themselves the results of painstaking bi/multilateral bargaining. Additionally, the six-parties will need to address what has transpired in North Korea between the time the talks ended and the time they convene in Beijing.

There may be breakdowns at every turn in the road to the denuclearization. There may be, for example, feuds over the degree of intrusiveness of inspections in the closed nation and over the scope and nature of any peaceful nuclear program the North is allowed to retain. There may be a divide within Pyongyang's new leadership and bureaucratic bickering may surface, hindering the process of denuclearization. However, it needs to be noted that the 9.19 and 2.13 accords contain everything that is needed to achieve a nuclear-free Korean peninsula: provision of energy (including LWR) for North Korea, normalization of its relations with the United States/Japan, a

---

[43] Ibid.

permanent peace regime replacing the armistice, and security cooperation in Northeast Asia. These measures will have to be combined in such a way that effectively induces North Korea into a grand settlement.

Diplomatic Normalization

In particular, it should be noted that the ace card the United States holds is US diplomatic normalization with the North. Although not a sufficient condition, it is a necessary one for the solution of the nuclear problem. In exchange for the North's dismantlement of all its nuclear facilities and removal of all its nuclear weapons and material, which is subject to international verification, the United States should normalize relations with Pyongyang. Then, the North should rejoin the NPT and immediately re-sign the safeguards agreement with the IAEA. Normalization of relations would bring about a solution to the missile-proliferation problem. In the normalization process, North Korea should be obligated to make a nonproliferation pledge and join the Missile Technology Control Regime.

The United States seems to view the normalization of its relations with North Korea as some sort of compensation to North Korea. In part, this perception was created by the North, which has considered normalization as critical evidence that the United States would no longer be hostile toward it and as an existential procedure that is absolutely necessary for survival. The United States could use this perception to its advantage when negotiating with the North. But, I stress, if it looks at the other side of this, Washington will find that the normalization will bring it great advantages in various dimensions. For example, due to a greatly alleviated threat perception, normalization could encourage North Korea to make a move toward more thorough reforms and give it confidence in opening up. After Sino-US normalization of relations, China became far more pragmatic and open, which suggests the possibility for a similar future for a postnormalization

North Korea.

Additionally, normalization will bring about expanded commercial and cultural exchanges that will inevitably lead to tremendous intellectual exchanges between North Korea and the West. As long as the North harbors intentions to build nuclear weapons, it is not possible to guarantee that the North will continue to be nonnuclear, given that it has the technology and the scientists necessary to build nuclear weapons. Therefore, it is essential to the United States that North Korea voluntarily abandons its intentions. In a post–Cold War period, it is not difficult to predict where the exchange of ideas between Socialism/Juche/Songun politics and market democracy will lead North Korea. A pragmatic, open, secure North Korea, integrated with the international society, would find that efforts to possess nuclear weapons would in fact harm its fundamental interests.

On nuclear verification, normalization of relations will lead to another advantage for the United States As is well known, outside verification of WMD, especially uranium-related, is difficult to accomplish mainly because such programs are clandestine. The more foreigners travel and stay in North Korea, the more opportunities arise for thorough inspection and verification on secret programs of dangerous and illicit weapons. It is difficult to imagine internationally integrated and open societies such as that of "democratized and open" South Korea could possess "secret" programs of WMD. In a similar manner, but more broadly understood, normalization will help the United States set up a sound North Korea policy that is not based on circumstantial evidence and hunches but on what is actually happening within the North. A North Korea policy of the United States that reflects reality will serve its other strategic interests in Northeast Asia, including stable management of Sino-US relations.

Skeptics express concern that "propping up" North Korea's economy will result in an economically and militarily stronger North

Korea that will become more emboldened to attack South Korea and US interests overseas. However, this may be true only if North Korea has nuclear weapons. North Korea understands that a war initiated by a denuclearized North would be an act of suicide. It knows that its military adventurism would be opposed by both China and Russia and instantly defeated by the mighty firepower of the combined forces of the United States and South Korea. The North has no reason to become militarily adventurous, once stripped of nuclear weapons. The normalization of relations would also greatly improve the human-rights situation in North Korea. The United States should be reminded, for example, that its efforts to promote human rights in South Korea and China were much more successful after a civil society had evolved in South Korea and a US embassy was established in China.

However, there is some pessimism and suspicion in both the United States and South Korea regarding the effectiveness of the normalization of relations with the North. It is argued that North Korea will never abandon its nuclear-weapons programs, even if the United States normalizes relations with it. Many in the Bush administration had this doubt, and there are those in both the Obama and Lee administrations who share this view. They tend to believe that the North thinks that it can "have its cake and eat it, too." Some suggest that the North's statement on January 17, 2009 proves this point:

> Even if the North Korea-US relationship becomes normalized diplomatically, as long as there is any possibility for the US nuclear threat to remain present, our status as a nuclear country won't change at all.... North Korea can go on existing without the normalization of relations with the United States but it cannot go on existing without its nuclear deterrent.[44]

---

[44] The statement by a spokesperson of the foreign ministry was carried in Chosen Shinbo, a pro-North Korea newspaper based in Japan. "DPRK Foreign Ministry

But, one should not take this in the literal sense of the words, considering the context in which it was said. First, the statement was a retort. When US Secretary of State-designate Hillary Clinton, during her confirmation hearing, stated that normalized relations will not be possible without complete and verifiable denuclearization of North Korea, the North angrily answered:

> This (US) statement is an expression of America's "big power" mentality that portrays the normalization as if it were its present to us. It is also a distortion of the fundamentals in the nuclear issue of the Korean Peninsula.

Second, the North's statement was more of an attention-seeking tactic, meant for the incoming US administration. The statement was issued four days before Obama's inauguration as president. And it came amid an impasse in the six-party talks that aimed to end the North's nuclear drive.

Third, it is important to realize that the statement needs to be qualified by its conditional nature. The remarks were "if and then" clauses, not a unilateral declaration. In November 2002, Kim sent Bush a letter saying: "If the United States recognizes our sovereignty and assures non-aggression, it is our view that we should be able to find a way to resolve the nuclear issue in compliance with the demands of a new century." Further, he declared: "If the United States makes a bold decision, we will respond accordingly."[45] Again, "contextomy,"

Spokesman Dismisses US Wrong Assertion," *Korean Central News Agency* (January 17, 2009), www.kcna.co.jp.

45 According to Robert Gard, Lt. Gen. (USA, ret.), in the fall of 2002, Donald Gregg, former ambassador to South Korea during the Bush I presidency, and Donald Oberdorfer, former *Washington Post* reporter, delivered a message from Kim Jong-il to the White House: "If the United States recognizes our sovereignty and assures non-aggression, it is our view that we should be able to find a way to resolve the nuclear issue.... If the United States makes a bold decision, we will respond

the practice of quoting out of context, is not only a logical fallacy and a false attribution but also a dangerous distortion. As late as September 27, 2011, the official position held by the North was that it was still committed to denuclearizing the peninsula through dialogue and negotiation.

Fourth, it is significant that the North continues to emphasize its "predetermined" and "holy" position on the nuclear issue. As Pak Kil Yon, vice-minister of the North's foreign ministry, stated publicly at the United Nations in 2011, North Korea held that the "denuclearization of the entire peninsula was President Kim Il-sung's last instruction and an ultimate goal" of North Korea.[46] People can judge for themselves whether the Foreign Ministry's 2009 statement is more or less important than the last instruction by Kim Il-sung, whose bloodline determines the current and, probably, the future politics in North Korea. It thus seems reasonable to suggest, although one can never know for sure, that the North Korean statement in January 2009 does not justify the claim that the North will never give up its nuclear-weapons program. Meanwhile, it needs to be noted that North Korea's domestic conditions will not justify such a claim. As said earlier, the year 2012 marks the centennial anniversary of the birth of Kim Il-sung, with North Korea gearing up to show that it is a "strong and prosperous nation." That slogan raised the ante for receiving food aid. The years after it are not likely to be a lot better in terms of the food supply. In fact, 2012's

---

accordingly." Not only did the administration ignore this and other overtures, but shortly thereafter it persuaded South Korea and Japan to join in stopping the delivery of fuel oil to North Korea as required by the 1994 Agreed Framework: Robert Gard, "Negotiating with North Korea on Its Nuclear Program"; http://armscontrolcenter.org/policy/northkorea/articles/negotiating_with_north_korea_on_its_nuclear_program/.

[46] Statement by Pak Kil Yon, vice-minister of foreign affairs and head of delegation of the Democratic People's Republic of Korea in general debate of the sixty-sixth session of the UN General Assembly, New York, September 27, 2011.

joint editorial in state newspapers called food shortages a "burning issue." China is expected to provide the North with food and energy to prop it up, but China's calculus dictates that the amount of aid to the North be just enough for it to survive, not less and not more. Kim Jong-un and the North Korean elite surely know that the economic stress is—and will become—a heavy political burden for them. Can they say, "Let them eat yellow cake?"[47] No! They know they cannot.

Furthermore, the fact that North Korea is already nuclear-capable is critical. The North will be able to reconstitute a nuclear program at some future date if deemed necessary. This means it can abandon the nuclear program now if the "price is right." As long as the US objective is to dismantle North Korea's current nuclear-weapons program rather than eliminating its future nuclear capability (including the knowledge, technology, and natural resources), this assessment remains valid. In fact, the knowledge and technology certainly can hardly be "unlearned." In any case, the United States knows that the option of eliminating the capability is not currently on the table.[48] Such a step would be much more·easily accomplished, for example, as a program of Cooperative Threat Reduction after the North reforms itself and opens it up to the outside world.

In sum, the claim that the North will never give up its nuclear-weapons program appears to have only limited validity. Of course, it is difficult to predict how North Korea will react, but an educated guess is that, in the longer term, the North is likely to cut a deal with the United States and others (probably within the framework of the six-party talks). The North and others know well that it is going to be

---

[47] "Let Them Eat Yellowcake: Even Mass Hunger May Not Drive North Korea into Bargaining Away Its Nuclear Capability," *Economist*, January 14, 2012.

[48] The 9.19 agreement stipulates that "the DPRK committed to abandoning all nuclear weapons and existing nuclear programs."

a long rough road ahead of them. But they will choose to take steps to travel that road together because their cost-benefit analyses and the structural constraint under which they operate will urge them to do so.

* * * * *

The progress North Korea has made in developing nuclear weapons constitutes a serious threat not only to South Korea and the United States but also to the entire world, including the North itself, however paradoxical that sounds. Theoretically, there are a number of ways to remove the North Korean nuclear weapons and capabilities. But when approached from the perspective of feasibility, the only viable alternative is via diplomacy. Over the past two decades, various diplomatic approaches have been adopted and implemented with varying degrees of success. It is difficult to precisely evaluate the performance of each approach because it is hard to isolate the net effects of North Korea policies. However, it is also undeniable that one policy has been relatively superior to another in terms of eliciting discernible cooperative behaviors from North Korea, and, by logical extension, it is plausible to suggest that policy is more likely to bring about a solution to the North Korean nuclear problem. As discussed above, the North Korea policies based on idealism, inflexible reciprocity, and hard power have been infertile or, in some cases, even dangerous. I have argued that in order to reverse the current hostile inter-Korean and North Korea–US relations there should be a paradigmatic change in South Korea/US outlooks as far as their approaches to the North are concerned. The core of the change would involve reintroducing the notions of flexibility, balance, and soft power to their policies. Embracing the necessity of this paradigmatic shift, I have proposed a policy framework that is pragmatic, strategic, and comprehensive, and several concrete policy alternatives that I believe can facilitate North Korean denuclearization. The essence of diplomacy is getting the other guy to do what you want and to make him to think that it was his idea. The SPC approach will do that.

Chapter 5

Can the North Korean Nuclear Crisis Be Resolved?

Siegfried S. Hecker

My area of expertise is in the nuclear arena and that is primarily what I will address. However, since it is not possible to solve the nuclear crisis without addressing the underlying fundamental political issues, I offer my views in this area at the outset, as a nonexpert.

I am optimistic in the long term but pessimistic in the short term that the nuclear crisis on the Korean peninsula can be resolved. The reason for my optimism in the long term is that Northeast Asia is the most dynamic, rapidly growing economic region of the world. North Korea is an island of instability in this sea of stability and growing prosperity. There is simply too much at stake for the surrounding countries to allow the nuclear crisis to destabilize the region and take it down into chaos. This situation is very different from Iran, the other current nuclear hotspot in the world. Iran represents a much greater global challenge because it is an island of instability in a sea of instability and turmoil; namely, the greater Middle East.

The reason for my pessimism about resolving the North Korean nuclear crisis in the short term is that it will take time for Washington and Pyongyang to overcome the mistrust built up by five decades of enmity and two decades of contentious nuclear diplomacy. Moreover, the nuclear crisis cannot be resolved in isolation. It is imperative to address the long-standing problems created by the division of the Korean peninsula and the remnants of the Korean War. Pyongyang developed its nuclear program to help assure regime survival and, today, appears to be using it to forge an outcome far beyond what Pyongyang could achieve economically or through conventional military means. Pyongyang has survived by playing its poor hand to perfection in large part because the countries working to denuclearize the Korean peninsula have differing and conflicting interests in North Korea and the region.

These differences, even if acknowledged, have not been resolved, and they prevent a unified strategy toward North Korea from being developed. South Korea has vacillated during the past two decades between sunshine and hard-line. China continues to stress peace and stability on the Korean peninsula, regardless of how confrontational Pyongyang becomes. Japan appears to have sidelined itself from serious influence by focusing almost exclusively on the abductees issue. Russia advocates a soft approach to Pyongyang formally, but behind the scenes blames Washington for exacerbating the crisis. And the United States continues to be fixated by the nuclear issue and has handcuffed itself with domestic politics, which prevents it from taking the kind of creative approach required to resolve the crisis.

The recent agreement by Washington and Pyongyang to renew diplomatic dialogue is a positive sign of progress. On February 29, 2012, North Korea agreed to implement a moratorium on long-range missile launches, nuclear tests, and nuclear activities at Yongbyon, including uranium-enrichment activities in return for nutritional aid from the United States.[1] The agreement is a small, but necessary, step

[1] Statement on "US-DPRK Bilateral Relations." US Department of State, February 29, 2012, Washington DC, and "DPRK Foreign Ministry Spokesman on Result of DPRK-US Talks," Korea Central News Agency (KCNA), February 29, 2012. Pyongyang, North Korea.

to prevent the nuclear crisis from getting worse as it did during the past three years. Below, I provide a brief history of Pyongyang's nuclear program, review nuclear developments during the past three years, offer some suggestions of how to proceed in the near term to reduce the nuclear crisis on the Korean peninsula and some long-term suggestions of how to resolve the nuclear crisis and attain the ultimate goal of the denuclearization of the peninsula.

A Brief History of the Nuclear Crisis

The North Korean nuclear crisis is often viewed in black-and-white terms—that is, Pyongyang either has or doesn't have nuclear weapons or it will or will not give up such weapons. However, any serious attempt at resolving the nuclear crisis must recognize that the details of the nuclear program matter. The North Korean nuclear program is particularly complex: it was begun more than fifty years ago ostensibly to provide the benefits of nuclear technologies for the people of North Korea, but also because Kim Il-sung wanted to build the technological foundation for nuclear weapons.[2]

Pyongyang has pursued both nuclear-energy and nuclear-weapons capabilities for five decades. Much of the underlying technological base is the same. Over time, Pyongyang's emphasis on one or the other has been influenced heavily by political considerations and diplomacy, particularly as it relates to Washington. However, no matter if it was bombs or electricity that Pyongyang favored, it always kept open the option for the other. Washington, on the other hand, has continued

---

[2] See Siegfried S. Hecker, "Lessons Learned from the North Korean Nuclear Crises," *Daedalus* (Winter 2010): 44–56 for a detailed treatment of how and why North Korea developed nuclear weapons.

to view North Korea strictly through a nuclear weapons lens; first stating that it will never allow Pyongyang to obtain nuclear weapons and, once that failed, proclaiming that it will not accept Pyongyang keeping them.

Kim Il-sung laid the foundation for nuclear-technology development in the early 1950s. The Soviet Peaceful Atom initiative, modeled after President Dwight Eisenhower's Atoms for Peace, enabled Soviet universities and nuclear-research centers to train several hundred North Korean students and researchers. The Soviets built a research reactor, the IRT-2000, and associated nuclear facilities at Yongbyon in the 1960s. North Korean specialists trained at these facilities, and by the 1970s they were prepared to launch a nuclear program without external assistance.

The North Koreans first used their newly acquired technical capabilities to construct an indigenous experimental gas-graphite reactor. North Korea's decision to build gas-cooled, graphite-moderated reactors was a logical choice at the time for an indigenous North Korean energy program because gas-graphite reactors can operate with natural uranium fuel and, hence, do not require enrichment of uranium.[3] Although North Korea may have experimented with enrichment technologies, commercial enrichment capabilities were beyond its reach and difficult to acquire.[4] North Korea's ambitious program began

---

[3] The gas-graphite reactors were patterned after the British Calder Hall Magnox reactor, whose technical specifications were readily available because they were widely disseminated in the United Kingdom.

[4] The alternative path for natural uranium-fueled reactors is a heavy-water reactor, such as the Canadian CANDU reactor. This was India's choice for its first reactor, which was constructed by Canada with US-supplied heavy water. However, after India used the plutonium produced by that reactor for its first nuclear test in 1974, it would have been difficult for North Korea to get external assistance. North Korea required external assistance because it did not have the capacity to produce heavy water.

with an experimental 5-megawatt electric (MWe) reactor, which became operational in 1986. Construction of that reactor was followed by a scaled-up 50-MWe reactor and a 200-MWe power reactor, although neither was ever completed.

North Korea quickly mastered all aspects of the gas-graphite-reactor fuel cycle. It built fuel fabrication facilities and a large-scale reprocessing facility, which enabled extraction of plutonium from spent fuel.[5] Unlike the Soviet-built research facilities, the new facilities were built and operated without being declared to or inspected by the International Atomic Energy Agency (IAEA). Pyongyang had no legal obligation to declare these facilities because it was not yet a member of the Nuclear Non-Proliferation Treaty (NPT). US reconnaissance satellites picked up signs of the reactor construction in the early 1980s and of the reprocessing facility in the late 1980s. It was not until 1989, when South Korea leaked US satellite data of the reprocessing facility, that the international community first became aware of and concerned about North Korea's indigenous nuclear program. The concern stems from the fact that gas-graphite reactors are capable of producing weapons-grade plutonium while generating electrical power and heat. So, whereas Pyongyang's choice of gas-graphite reactors for its energy program was logical, it was also the best choice to develop a nuclear-weapons option.

In parallel, North Korea asked the Soviets to build light-water reactors (LWRs) to help meet North Korea's energy demands. North Korea joined the NPT in 1985 because the Soviets made consideration of constructing LWRs contingent upon joining the treaty. These reactors, however, never materialized because of the demise of the Soviet Union. Pyongyang kept inspectors out of its new facilities until

---

5 The reprocessing facility resembles an extension of the design of the Eurochem reprocessing plant in Belgium.

1992, by which time it had all of the pieces in place for the plutonium-fuel cycle. This move coincided with several diplomatic initiatives and President George H. W. Bush's decision to withdraw all US nuclear weapons from South Korea. By this time, the 5-MWe experimental reactor produced electricity and heat for the town of Yongbyong, as well as approximately 6 kilograms (roughly one bomb's worth) of weapons-grade plutonium per year. The fuel fabrication and reprocessing facilities were operational, and the two bigger gas-graphite reactors were under construction.

In 1992, Pyongyang opened the window on its nuclear program for diplomatic reasons explained below, but closed it quickly when IAEA inspectors uncovered discrepancies between their own nuclear measurements at Yongbyon and Pyongyang's declaration. Pyongyang responded to IAEA accusations by announcing its intent to withdraw from the NPT. Negotiations started in June 1993, but stalemated. In 1994, when North Korea unloaded the reactor's fuel containing an estimated 20 to 30 kilograms of plutonium, Washington and Pyongyang came close to war before former President Jimmy Carter intervened and brokered a freeze.

Intense negotiations in Geneva led to the Agreed Framework,[6] which changed North Korea's nuclear-technical trajectory dramatically. Pyongyang agreed to give up its indigenous gas-graphite reactor program for the promise of two LWRs, to be supplied by the United States, South Korea, and Japan. The spent fuel rods unloaded from

---

[6] The Agreed Framework signed between the United States and North Korea on October 21, 1994, in Geneva agreed to have North Korea freeze its existing nuclear program. In addition to US supply of LWRs and delivery of heavy fuel oil, the two sides agreed to move toward full normalization of political and economic relations and work together for peace and security on a nuclear-free Korean peninsula. See Joel S. Wit, Daniel B. Poneman, and Robert L. Gallucci, *Going Critical: The First North Korean Nuclear Crisis* (Washington, DC: Brookings Institution Press, 2004) for informative discussions of the Agreed Framework and North Korean crisis in the 1990s.

the 5-MWe reactor were repackaged by a US technical team and stored in the cooling pool for eventual removal from North Korea. Operation of the 5-MWe reactor, the fuel fabrication plant, and the reprocessing facility was halted and monitored by IAEA inspectors per special arrangement under the Agreed Framework. Construction of the two larger reactors was stopped.

Although Pyongyang halted its plutonium program during the Agreed Framework, it continued to expand its missile program, including conducting a long-range rocket launch over Japan in 1998. It also explored uranium enrichment.[7] During its first formal encounter with Pyongyang in October 2002, the Bush administration, which was adamantly opposed to the Agreed Framework, accused Pyongyang of covertly pursuing the alternative High Enriched Uranium (HEU) path to the bomb. This altercation effectively ended the Agreed Framework and changed Pyongyang's technical and political trajectory again. After years of denying any uranium enrichment activities, Pyongyang showed my Stanford University colleagues and me a modern uranium centrifuge enrichment plant in November 2010.[8] It is now clear that Pyongyang was pursuing an enrichment program in the late 1990s and that the US intelligence estimate was correct. Nevertheless, the Bush administration's political confrontation backfired. Washington traded a threat that would have taken years to materialize into nuclear

---

[7] In the late 1990s, Pyongyang is reported to have acquired centrifuge technology from Pakistan's A. Q. Khan, as reported by Pervez Musharraf in his book *In the Line of Fire: A Memoir* (New York: Free Press, 2006). Additional evidence, including the purchase of aluminum tubes suitable for centrifuge rotors from Russia and attempted purchase from Germany, is discussed in Hui Zhang, "Assessing North Korea's Uranium Enrichment Capabilities," *Bulletin of the Atomic Scientists* (June 18, 2009); http://www.thebulletin.org/web-edition/features/assessing-north-koreas-uranium-enrichment-capabilities.

[8] Siegfried S. Hecker (2010) A Return Trip to North Korea's Yongbyon Nuclear Complex. *Center for International Security and Cooperation*, Stanford University, November 20. Available at: http://iis-db.stanford.edu/pubs/23035/HeckerYongbyon.pdf.

weapons for one that took less than one year.

Following the altercation with the Bush Administration in late 2002, North Korea became the first nation to withdraw from the NPT.[9] It expelled international inspectors and announced that it would strengthen its nuclear deterrent. By the end of 2003, which also marked the invasion of Iraq and the fall of Saddam Hussein, Pyongyang was eager to have Washington believe it had the bomb. It used my first trip to North Korea, an unofficial, Track II trip led by my Stanford University colleague John W. Lewis, to send that message back to Washington. In a carefully choreographed tour of the Yongbyon nuclear complex in January 2004, Pyongyang gave me remarkable access to nuclear facilities and nuclear scientists and allowed me to hold 200 grams of plutonium bomb fuel (in a sealed glass jar), all to convince me it had a "deterrent."[10]

Over the next five years, Pyongyang built and demonstrated its nuclear-weapons capabilities while it engaged off and on in the six-party talks, which it joined under pressure from the Chinese.[11] We do not know exactly when Pyongyang got its first bomb, but we know it has made significant strides since 2003. Since the 5-MWe reactor was restarted in 2003, it operated for approximately three years before it was shut down again (July 2007). North Korea has conducted three reprocessing campaigns since 2003. The reprocessed plutonium, combined with the roughly 2 to 10 kilograms North Korea may have produced before 1994, yields an estimated plutonium production of 40 to 60 kilograms, of which we estimate 24 to 42 kilograms are available for weaponization today.

---

9 Hecker, "Lessons Learned," p. 47.

10 Siegfried S. Hecker, "The Nuclear Crisis in North Korea," *The Bridge*, National Academy of Engineering, 17–23 (Summer 2004).

11 The six-party talks, which were initiated in 2003, involved the United States, North Korea, and its four neighbors: South Korea, China, Japan, and Russia.

North Korea also conducted two nuclear tests of plutonium devices. Its first nuclear test in October 2006 was only partially successful: its yield was estimated at less than one kiloton. Initial estimates of the yield of the second test, which occurred in 2009, were 2 to 4 kilotons. This estimate has been revised upward to 4.6 kilotons in a recent analysis.[12] Consequently, the North may be able to design a Nagasaki-like bomb with a yield of up to 20 kilotons, but delivery is likely to be restricted to aircraft, boat, or van. For North Korea to gain enough confidence to mount a miniaturized design on a missile, it will have to test again. Hence, the technical and military driving forces for additional tests are high, even if, as Pyongyang surely knows, the political risks are also high.

The six-party diplomatic process has so far failed to do much to constrain Pyongyang's nuclear ambitions. It resulted in shutting down the 5-MWe plutonium production reactor, but did not curtail its drive for uranium-enrichment capabilities; nor did it appear to stop its missile or nuclear exports. Not surprisingly, many observers now look at the last two decades as a dismal diplomatic failure because Pyongyang's nuclear program was not eliminated. It is important, however, take a closer look at what Pyongyang actually achieved technically—or, perhaps more importantly, what it did not achieve. It failed to get commercial nuclear power. Although Pyongyang now has nuclear weapons, its weapons program is much smaller than it

---

[12] J. R. Murphy, B. C. Kohl, J. L. Stevens, T. J. Bennet, and H. G. Israelsson, "Exploitation of the IMS and Other Data for a Compressive Advanced Analysis of the North Korean Nuclear Test," *2010 Monitoring Research Review: Ground-Based Nuclear Explosion Monitoring Technology*, Science Applications International Corporation, pp. 456–465. Another recent analysis claims a minimum yield of 5.7 kilotons (E. Rougier, H. J. Patton, E. E. Knight, and C. R. Bradley, "Constraints on Burial Depth and Yield of the 25 May 2009 North Korean Test from Hydrodynamic Simulations in a Granite Medium," *Geophysical Research Letters*, vol. 38, L16316 (August 30, 2011).

would have been if left unchecked. With the capabilities it already had or was soon to complete by the early 1990s, Pyongyang today could have an arsenal of a hundred or more nuclear weapons. Instead, it has enough only enough plutonium for four to eight weapons and currently is not producing more. North Korea has the capacity to put the 5-MWe reactor back into operation and produce one bomb's worth of plutonium annually for the foreseeable future, but it has not taken steps to do so, perhaps indicating that it believes its small nuclear arsenal provides a sufficient nuclear deterrent. It apparently has a modern centrifuge facility, but it is not clear how much, if any, highly enriched uranium may have been produced for use as an alternate bomb fuel.

Pyongyang's export of missiles and nuclear technologies appears not to have been constrained. Pyongyang has pursued an extensive missile program for decades. It built its initial capability, obtained from the Soviets, into a formidable short-range missile force and developed an ambitious export business for reengineered Soviet missiles. Its principal customers have been Pakistan, Iran, Syria, Libya, Egypt, and Burma. Pyongyang's long-range missile development has been slow and not a great technical success. After the 1998 launch, it delayed its second launch until July 2006, primarily because of the missile moratorium it declared in 1999. However, the second launch failed instantly when the rocket apparently hit the gantry. Its third test, in April 2009, successfully lifted the first two stages over the Pacific, but the third stage failed.

We have much less information about its nuclear exports. However, evidence is overwhelming that Pyongyang built a plutonium-producing reactor for Syria that was destroyed by an Israeli air raid in September 2007. It appears quite likely that it exported to Libya uranium hexafluoride, the precursor to highly enriched uranium

(HEU). There are also grounds to suspect nuclear cooperation with Pakistan and Burma.[13] Cooperation with Iran is the greatest concern because Iran is putting in place all of the pieces for a nuclear-weapons option, and its nuclear capabilities complement those of North Korea.[14]

Recent Nuclear Developments, 2010–2012

The end of 2011 marked the third year of diplomatic standoff between North Korea and the United States. North Korea's third long-range missile launch in 2009 resulted in anticipated United Nations condemnation, which, in turn, triggered Pyongyang's expulsion of the international inspectors from its Yongbyon nuclear complex, followed by withdrawal from the six-party talks and by a second nuclear test. The year 2010 was marked by clashes and a dangerous spike in tensions on the Korean peninsula; namely, the sinking of the South Korean corvette, the *Cheonan*, and the North Korean shelling of Yeongpyeon Island. In contrast, 2011 was a year of diplomatic calm and rapprochement up to the death of Kim Jong-il and the leadership transition to his young son, Kim Jong-un.

---

[13] The evidence for North Korean assistance to Syria is strong; see David Albright and Paul Brannan, "The Al Kibar Reactor: Extraordinary Camouflage, Troubling Implications," Institute for Science and International Security (ISIS) Report, May 12, 2008; http://isis-online.org/publications/syria/index.html. Evidence of cooperation with Libya is less conclusive, yet cooperation likely did take place; see David E. Sanger and William J. Broad, "Tests Said to Tie Deal on Uranium to North Korea," *New York Times*, February 2, 2005. Evidence of nuclear cooperation with Burma is weak, but possible; see Julian Borger, "Burma Suspected of Forming Nuclear Link with North Korea," July 21, 2009; http://www.guardian.co.uk/world/2009/jul/21/burma-north-korea-nuclear-clinton.

[14] Siegfried S. Hecker and William Liou, "Dangerous Dealings: North Korea's Nuclear Capabilities and the Threat of Export to Iran," *Arms Control Today* 37, no. 2 (2007); http://www.armscontrol.org/act/2007_03/heckerliou; and Siegfried S. Hecker, "From Pyongyang to Tehran, with nukes," op-ed, *Foreign Policy* (May 26, 2009).

With the transition barely under way, it was surprising that in late February North Korea agreed to implement a moratorium on long-range missile launches, nuclear tests, and nuclear activities at Yongbyon, including uranium-enrichment activities, in return for US nutritional aid.[15] Washington and Pyongyang issued independent statements describing the agreement. The statements differed slightly, but significantly they exposed important unresolved issues that must be addressed before each party will agree to return to the six-party negotiating table. Pyongyang also agreed to allow IAEA inspectors to monitor the moratorium on uranium-enrichment activities at Yongbyon.[16] This agreement was necessary to prevent the nuclear crisis in North Korea from getting worse. I will first describe the troubling nuclear developments in the past two years before offering some suggestions on how to proceed toward the goal of the denuclearization of the Korean peninsula.

During the diplomatic calm of 2011, Pyongyang revealed very little about its nuclear progress, but all signs pointed to a continuing march toward a more threatening nuclear-weapons capability. The most alarming development was a combination of two things: the operation of a modern uranium centrifuge facility and the presence of road-mobile intermediate-range ballistic missiles (IRBMs). An added source of stress was North Korea's likelihood of cooperating with illicit nuclear programs of other countries, such as Iran, and the likelihood of it importing and exporting nuclear technologies to expand its own programs or aid those of others.

---

[15] Statement on "US-DPRK Bilateral Relations," and "DPRK Foreign Ministry Spokesman on Result of DPRK-US Talks."

[16] The American statement includes the language "and confirm the disablement of the 5-MWe reactor and associated facilities, whereas the North Koreans state "while productive dialogues continue." The North Korean statement also adds "Once the six-party talks are resumed, priority will be given to the discussion of issues concerning the lifting of sanctions on the DPRK and provision of light water reactors." The American statement does not include these provisions.

Light-Water Reactor

In 2010, North Korea greeted the world with the announcement that they would build an indigenous LWR and enrich uranium domestically to fuel it.[17] During my 2010 visit to Yongbyon with Stanford colleagues John Lewis and Robert Carlin, we confirmed that construction had begun on a 25–30 megawatt electric (MWe) LWR and that the North Koreans had built a modern, sophisticated uranium-enrichment facility, ostensibly to provide fuel for the reactor. Commercially available satellite imagery allows us to track the progress of North Korea's LWR construction from September 26, 2010 (just prior to our visit) to February 3, 2012—as shown in Figure 1. Early images indicate that the construction of this new LWR began in late September 2010, near the site of the destroyed cooling tower of the 5-MWe gas-graphite reactor. One of the latest available close-up overhead images taken on February 3 (Figure 2) shows that many of the reactor's external structures are almost complete. Much progress has been made on the turbine generator hall; a traveling crane rail was installed and the roof just completed.

Using overhead images from Figure 2, a 3-D model (Figure 3) of the LWR was constructed.[18] Our analysis confirms Pyongyang's plan to use this experimental reactor for electricity production. The rapid progress of construction also demonstrates that North Korea still has impressive manufacturing capabilities, in spite of the last two decades of economic downturn. However, we view this progress with alarm given the safety concerns associated with building and operating

---

[17] "N.Korea to Build Light Water Reactor Soon: State Media," Agence France-Presse, March 29, 2010. The North Korean's MOFA announced that they would build an indigenous LWR in April 2009 as well, but the 2010 pronouncement was considered more significant.

[18] See N. Milonopoulos, S. Hecker, and R. Carlin, "North Korea from 30,000 feet," *Bulletin of the Atomic Scientists*, January 6, 2012; http://thebulletin.org/web-edition/features/north-korea-30000-feet.

an LWR. What is especially troubling is that this is a new endeavor for North Korea, and its technical specialists have not been part of the global nuclear-safety community. These concerns were exacerbated by the nuclear accident at the Fukushima Daiichi Nuclear Power Plant in March 2011.

Uranium Enrichment

During our November 2010 visit to Yongbyon, we were stunned to find a newly constructed, modern 2,000-centrifuge uranium-enrichment plant.[19] Our visit answered some questions regarding the North's nuclear facilities at Yongbyon, but it raised many more. Though the Yongbyon uranium centrifuge enrichment facility looked complete, we were unable to verify that it was operational. We were told that the facility was producing low-enriched uranium (LEU) destined for the small, experimental LWR under construction; the facility appeared to house sophisticated centrifuges (apparently of the P-2 or G-2 design) and was sized properly for production of fuel for the experimental LWR.[20] We were also told that the facility became operational just a few days before our arrival, but we were unable to corroborate that. Whatever its status during our visit, the facility may be fully operational now, more than a year later. They may also have run into the typical difficulties of getting centrifuge cascades to operate smoothly, so we do not know how much LEU has been produced to date, what

---

[19] Siegfried S. Hecker, "A Return Trip to North Korea's Yongbyon Nuclear Complex," *Center for International Security and Cooperation*, Stanford University, November 20, 2010; http://iis-db.stanford.edu/pubs/23035/HeckerYongbyon.pdf.

[20] Yongbyon officials claimed annual throughput capacity was 8,000 separative work units (the measurement of the separation during the enrichment process), indicating that the centrifuges were second generation, or so-called P-2s; first generation centrifuges, by comparison, produce an annual throughput capacity of about 2,000 separative work units (Hecker, "Return Trip," 2010).

the current production rates are, or what is the disposition of any LEU that has been produced to date. Although the facility is likely producing LEU for the reactor, it could conceivably be producing HEU bomb fuel. And if it is configured to do this, the reactor could produce roughly 40 kilograms of HEU annually, enough for at least one bomb per year.

As far as I know, no foreigners have been given access to the facility since our November 2010 visit. Our requests for a return visit have so far been denied. By tracking the construction of the Yongbyon centrifuge plant via overhead imagery, however, we conclude that North Korea must also have an undisclosed, pilot-scale centrifuge facility elsewhere to have made it possible for such rapid installation progress at Yongbyon. Figure 4 is a rough schematic of the interior layout of Building 4, which was renovated from the former fuel-rod fabrication plant to a centrifuge hall. In the November 2010 visit, we observed approximately 2,000 centrifuges, divided into six cascades, from the second-floor observation platform identified in the diagram. Figure 5 represents a 3-D model of the exterior of the Building 4 renovated centrifuge building. Unless Pyongyang allows access to this facility as part of the new agreement, however, the world will not know if it is fully operational or much else about its sophistication or that of other undisclosed centrifuge facilities.

The Yongbyon centrifuge facility could not have been constructed from scratch and made operational in only 18 months, between April 2009 and November 2010, as Pyongyang has claimed. It is likely that the North had one full cascade (about 340 centrifuges) operational at a separate site long before it moved into the renovated Yongbyon fuel-fabrication building and revealed their centrifuge program in November 2010. The size of any clandestine program is likely constrained, but constrained or not, North Korea may be producing some HEU now, and may have been doing so for some time.

Plutonium

The Yongbyon plutonium facilities remained dormant in 2011. The 5-MWe plutonium production reactor, operational since 1986, was shut down in July 2007 and has not been restarted. The reprocessing facility ceased operations in 2009, but remains on stand-by status. Presently, North Korea is not producing any plutonium and there is no plutonium in the pipeline. The key facilities could be reactivated if necessary; it would take approximately six months to do so.[21] Therefore, my estimate that North Korea has roughly 24 to 42 kilograms (approximately four to eight bombs worth) of plutonium still holds today.

The decision to pursue uranium-enrichment instead of plutonium production is puzzling—if Pyongyang simply wants to make more bomb fuel.[22] The missing piece of the puzzle, however, is that Pyongyang has long sought LWRs for electricity production, first from the Soviet Union and then from the United States[23]—and this type of reactor fuel requires enrichment, which, in turn, opens the door to the weapons option since the centrifuge facilities needed for the reactors can also be converted to produce weapon-grade HEU bomb fuel. Thus, choosing the uranium route provides Pyongyang with a viable dual-track option: LEU for nuclear electricity with LWRs and HEU for the second route to the bomb to augment its small plutonium bomb inventory.

---

[21] If reactivated, the reactor is capable of producing only six kilograms of plutonium, roughly one bomb's worth, per year.

[22] Plutonium is used in all states with nuclear weapons. China switched from HEU to plutonium early in its program, and Pakistan has begun to employ plutonium in addition to HEU.

[23] See Siegfried S. Hecker, Chaim Braun, and Robert L. Carlin, "North Korea's Light-Water Reactor Ambitions," *Journal of Nuclear Materials Management* 39, no. 10 (Spring 2011).

Nuclear Weapons and Delivery Systems

Pyongyang has the bomb, but not much of a nuclear arsenal. We have reasonable confidence in the number of bombs—roughly four to eight—because plutonium inventories are easy to assess, but we simply don't know their sophistication. Since it has shut down its plutonium facilities, Pyongyang apparently is not planning to increase the number of plutonium bombs significantly.

We assume the North is working on missile-capable nuclear systems, but employing miniaturized nuclear warheads will require more nuclear tests. Satellite imagery captured in 2011 showed preparations for what possibly could be another test tunnel, which is located near the two previous tests in the Kilju region.[24] Hence, the nuclear-test moratorium agreed to in the February statements is particularly important in order to halt Pyongyang's drive toward more sophisticated, smaller nuclear weapons that could be mounted on their missiles. If Pyongyang breaks the test moratorium, it will almost certainly be a test of a miniaturized design, but we don't know if it will be with plutonium or HEU.

In October 2010, North Korea publicly exhibited, for the first time, a road-mobile intermediate-range ballistic missile at a military parade in Pyongyang. Dubbed the Musudan by US intelligence services, the IRBM can travel an estimated 3,000 to 5,000 kilometers, apparently farther than any other missile in the North Korean arsenal.[25] Although never flight-tested, the missile represents a big step forward for Pyongyang because it is road mobile and, hence, difficult to find. In June 2011, the then US defense secretary, Robert Gates, expressed

---

[24] "North Korea Prepared to Detonate Third Nuclear Device, South Says," *Global Security Newswire*, April 19, 2011; http://gsn.nti.org/gsn/nw_20110419_5975.php.

[25] "North Korea Rolls Out Ballistic Missiles," *Global Security Newswire*, October 13, 2010; http://gsn.nti.org/gsn/nw_20101013_1452.php.

concern that North Korea had also been developing a road-mobile intercontinental ballistic missile (ICBM).[26]

Nuclear Cooperation, Imports, and Exports

Though 2011 cast little light on North Korea's nuclear relationships with outside countries, there is certainly cause for increased concern, considering developments like the Musudan and the new uranium-enrichment facilities at Yongbyon. North Korea has a history of being a quick study (it became self-sufficient for the entire plutonium fuel cycle after initial help from the Soviets) and enterprising (Pyongyang almost certainly built a plutonium production reactor for Syria).[27]

North Korea has historically relied on importing key materials and components for its uranium-centrifuge program. I believe it still does not have all the requisite capabilities today and has to rely on imports to expand its program. The centrifuge facility we were shown in 2010 apparently benefited from imports from Europe, Russia, Japan, and the A. Q. Khan network prior to 2003. It also profited from training and consultation with Pakistani enrichment specialists in the Khan Research Laboratories (KRL).[28] Today, the most likely acquisition route for key materials and components is through China.[29]

---

[26] In June 2011, US Secretary of Defense Gates noted: "With the continued development of long-range missiles and potentially a road-mobile intercontinental ballistic missile and their continued development of nuclear weapons, North Korea is in the process of becoming a direct threat to the United States"; http://www.iiss.org/conferences/the-shangri-la-dialogue/shangri-la-dialogue-2011/speeches/first-plenary-session/qa/. Also available at http://www.thedailybeast.com/articles/2011/06/21/robert-gates-interview-his-lingering-concerns-about-u-s-supremacy-nuclear-proliferation-and-more.html.

[27] There is little chance that North Korea has done this anywhere else. Additionally, reactors are difficult to hide and are vulnerable to foreign intervention, as was demonstrated by Israel's destruction of the Syrian reactor in 2007.

[28] Hecker, Braun, and Carlin, "North Korea's Light-Water Reactor Ambitions," pp. 18–25.

[29] David Albright and Paul Brannan, *Taking Stock: North Korea's Uranium Enrichment Program*, Institute for Science and International Security, October 8, 2010.

Over the past ten years, North Korea has developed a uranium-export business, supplying Libya with 1.8 metric tons of uranium hexafluoride before Muammar Qaddafi terminated the program in 2003.[30] The reactor built for Syria by North Korea at Dair Alzour also would have provided a lucrative fuel-export business for North Korea had it not been bombed by Israel in 2007.[31] In 2011, Pyongyang may have continued to export nuclear technologies, knowhow, and precursor materials like uranium hexafluoride or, potentially, HEU itself to dangerous states. None of these are easy to detect or easy to stop. The footprint for uranium-centrifuge activities is small, detection is difficult, and Pyongyang could claim exports are for civilian applications.

Virtually all North Korean missiles are copies or derivatives of Soviet missiles. More than 20 years ago, Pyongyang turned from import to export, becoming the major supplier of missiles and the means to manufacture them to the most unstable parts of the world. Their export business has slowed down considerably,[32] but it appears that North Korea is now collaborating closely with Iran's missile establishment.[33] Nuclear cooperation between North Korea and Iran, including the export and import of sensitive nuclear and missile technology, could greatly benefit both countries—reactor, plutonium, and weapons tech-

---

30 IAEA Board of Governors General Conference, "Application of Safeguards in the Democratic People's Republic of Korea," GOV/2011/53-GC(55)/24, September 2, 2011, p. 10, sec. 50.

31 IAEA Board of Governors, "Implementation of NPT Safeguards Agreement in the Syrian Arab Republic," GOV/2011/30, 2 May 2011, pp. 1-9; http://www.iaea.org/Publications/Documents/Board/2011/gov2011-30.pdf.

32 Joshua Pollack, "Ballistic Trajectory: The Evolution of North Korea's Ballistic Missile Market," *Nonproliferation Review* 18, no. 2 (2011): 411–429.

33 See *Iran's Nuclear and Missile Potential: A Joint Threat Assessment by US and Russian Technical Experts* (East-West Institute, May 2009); http://docs.ewi.info/JTA.pdf. Also see Mark Fitzpatrick, ed. *North Korean Security Challenges: A Net Assessment*, an IISS strategic dossier (London: International Institute for Strategic Studies, July 2011): pp. 179–182.

nologies from North Korea to Iran; centrifuge technologies and missile technologies in both directions.[34]

Prospects for Resolution of the North Korean Nuclear Crisis

In the Short Term

The centrifuge revelations in 2010 complicated an already gridlocked diplomatic six-party process, particularly reinforcing the hardliners' stance in Washington and Seoul against diplomatic engagement with Pyongyang. At the same time, events in Libya reinforced Pyongyang's conviction that ceding ground on the nuclear front is dangerous and possibly fatal.

Despite these negative developments, in February 2012 North Korea and the United States agreed on a moratorium on nuclear and missile testing and suspension of the enrichment plant in Yongbyon. North Korean–US negotiating history, and the fact that the two sides issued statements with somewhat different language, suggest that this new negotiation will be difficult and drawn out. Both sides reaffirmed their commitment to the September 19, 2005 Joint Statement (for realizing the denuclearization of the entire Korean peninsula) and recognized that the 1953 Armistice Agreement is the cornerstone of peace and stability on the Korean peninsula until the conclusion of a peace treaty. But they have different views of what denuclearization means and what will be required to verify and monitor it.

In the short term, preventing Pyongyang from substantially

---

[34] These concerns, which I previously expressed in 2009 (*Foreign Policy*), have been reinforced by one of the recent IAEA reports ("Implementation," 2011).

enhancing its nuclear program is the best that we can hope to achieve. Denuclearization remains the long-term goal; but the six parties must be prepared to address broader political, security, and economic issues in North Korea and Northeast Asia.

Although the US and North Korean statements have significantly different language reflecting their different expectations, the basic bargain—a nuclear moratorium in exchange for nutritional aid—should allow them to get back to the six-party table and begin to take meaningful steps to reduce the nuclear risks and address Pyongyang's broader concerns. Below, I outline some potential steps that should be considered early in the next round of negotiations.

Uranium Enrichment

While I believe the existence of a clandestine uranium-enrichment facility is highly likely, I do not know how large and how sophisticated such a facility (or facilities) may be, and if these facilities are producing LEU or HEU or both. This undeclared facility is unlikely, however, to house a large number of centrifuges, because I believe North Korea must still import some of the key materials and components (such as high-strength maraging steel and aluminum alloys) in order to construct large numbers of centrifuges. Since it is highly unlikely that Pyongyang will admit the existence of such facilities (and even less likely to allow access), the most important near-term step is for the IAEA or a US technical team to get access to the known Yongbyon centrifuge facility. Access will likely be a major point of contention in the next round of discussions; Washington will argue for full access and intrusive inspection, Pyongyang will likely try to keep inspectors stationed outside the centrifuge hall, with no access, claiming that it, like all other centrifuge plant operators in the world, has a right to protect its proprietary technologies. Washington should press to get access at least as intrusive as what the IAEA currently has at other centrifuge facilities it inspects, including those at the Iranian plant in Natanz and the Brazilian enrichment plant in Resende.

The greatest risk posed by North Korea's centrifuge program is that it may be configured to produce large quantities of HEU that could greatly increase the number of nuclear weapons it could field. I do not believe this is currently possible. Highly enriched uranium does offer the simplest and most assured route to a primitive bomb—the so-called gun assembly used for the Hiroshima device. However, Pyongyang has already demonstrated a simple plutonium bomb, so there is no obvious benefit for North Korea to produce a simple HEU bomb, unless they build many of them (which they cannot do with plutonium since no more plutonium is currently being produced in North Korea). Using HEU can offer some benefits, including its use for building a more sophisticated, miniaturized nuclear warhead, but such a design would have to be tested before Pyongyang could have any confidence to mount one on a missile. Information on such design(s) was distributed to various clients of the A. Q. Khan network and may also be available to North Korea.

It is crucial, therefore, to stop Pyongyang from importing large quantities of key centrifuge materials and components in order to prevent North Korea from building large additional centrifuge facilities now that it has apparently mastered the art of manufacturing and operating such facilities. Whereas previous supply routes for such materials and components likely originated in Europe and Russia or via the A. Q. Khan network in Pakistan and Malaysia, the most likely routes today go through the enormous and ill-controlled Chinese industrial sector. It should be our first order of discussion with China to help shut down such potential routes. Additionally, as part of the moratorium process, Washington should also attempt to place all LEU that may already have been produced at Yongbyon under IAEA safeguards to prevent potential subsequent conversion to HEU in undeclared North Korean enrichment facilities.

The Light-Water Reactor

There is no mention in the February statements of the fate of the experimental LWR being constructed at Yongbyon. The North Korean statement hints at the potential prospect that it may be willing to abandon that project since it asks for the provision of an LWR, which indicates Pyongyang is willing to get assistance and financial support in constructing a new LWR. The Agreed Framework had provisions for two modern, commercial LWRs of 1,000 MWe (compared with the Yongbyon experimental LWR designed for 25 to 30 MWe). The LWR issue need not be addressed at this time. Even if North Korea continues with the construction of the experimental LWR, it will take at least a couple of additional years before it can become fully operational. Moreover, as long as operation of the Yongbyon centrifuge plant remains suspended, the reactor cannot be fueled because it is highly unlikely that sufficient fuel has already been produced or that a clandestine facility has the capacity for sufficient LEU fuel.

In the longer term, it could well be that a provision of a modern, moderately-sized LWR may be the best way of keeping Pyongyang from completing and operating its own reactor with questionable safety credentials. An LWR reactor could be provided with guaranteed fuel supply and take-back arrangements. This, in turn, would eliminate the need for North Korea to have its own enrichment and reprocessing facilities, thereby removing the primary proliferation concerns. Pyongyang was prepared to accept this kind of arrangement a few years ago. However, it may be more difficult to get Pyongyang to agree to give up its modern centrifuge plant now that it has shown the world that it can build one.

Currently, it is politically impossible to consider providing an

LWR unless it is part of a bargain in which Pyongyang agrees to give up its nuclear weapons and consents to intrusive monitoring and verification mechanisms. If such a bargain can be worked out, Washington and its partners should consider revisiting the KEDO arrangement.[35] Alternatively, either China or Russia could supply a moderate-size LWR for North Korea. Another possibility for long-term consideration in a weapon-free Korean peninsula is to help Pyongyang convert its HEU-fueled IRT-2000 research reactor to LEU fuel to produce medical isotopes that North Korea's medical community could use and/or sell to neighboring countries. An even better solution may be to have South Korea build a version of its world-class research and isotope production reactor, the HANARO Reactor, for North Korea to produce medical isotopes. Seoul has just signed a deal to build such a reactor for Jordan.

The Yongbyon Plutonium-Production Complex

Our 2010 visit to Yongbyon confirmed that the 5-MWe reactor and the reprocessing facility were not operational and that a key building in the fuel-fabrication facility used for fuel-rod fabrication for that reactor was converted into the centrifuge hall. The chief engineer told us that these facilities are on stand-by, but they could be reactivated.[36] One of the first orders of a return to six-party negotiations should be to have Pyongyang take the steps necessary to permanently shut these facilities. These steps should involve permanently disabling the core

---

35 The Agreed Framework, negotiated in three substantive sessions (July 1993, August 1994, and September/October 1994), produced an agreement to construct two modern 1000-MWe LWRs in North Korea as part of a consortium eventually known as the Korean Peninsula Energy Development Organization (KEDO).

36 Siegfried S. Hecker, "A Return Trip to North Korea's Yongbyon Nuclear Complex," *Center for International Security and Cooperation*, Stanford University, November 20, 2010; http://iis-db.stanford.edu/pubs/23035/HeckerYongbyon.pdf.

of the reactor (that can be done by a number of means), selling the roughly 14,000 natural-uranium metal fuel rods (which make up more than one full core load of new fuel) and dismantling the front end of the plutonium reprocessing facility so that it would no longer be able to accept spent fuel. These actions will effectively and irreversibly shut Yongbyon's plutonium production facilities. To date, the facilities have been frozen several times, but never permanently taken out of commission. Eventually, Pyongyang must address the decommissioning and decontamination of the buildings and site.

Nuclear Weapons and Delivery Systems

The nuclear test and long-range missile launch moratorium was the most critical step to take. Another positive step would be for Pyongyang to agree to permanently destroy the test tunnel that has been prepared for a third nuclear test and completely abandon the test site. It is also imperative that North Korea not share its nuclear testing facilities or data with Iran. In due time, a complete declaration of fissile materials and nuclear-weapons inventories must be put on the negotiating table. In the longer term, Pyongyang must declare its entire nuclear program and be willing to have its dismantlement monitored and verified. Likewise, an accounting of the number of Musudan road-mobile missiles and other long-range missiles must be addressed.

Nuclear Exports and Imports

North Korea's exports of nuclear technologies or fissile materials can destabilize other parts of the world, like the Middle East. In our discussions with Pyongyang diplomats, we were told that they are prepared to discuss halting all "horizontal and vertical" proliferation—meaning no export and limiting the size of their own arsenal. The export issue must be addressed early on once negotiations resume. It will be one of the most difficult concessions to achieve because it is difficult to verify, and shutting down exports potentially represents a significant loss income of foreign currency for the regime.

Although I expect the negotiations to be difficult and drawn out, the process has begun. At this point, the most important first steps have been taken by the new Kim regime with the February 29, 2012, agreement. It is difficult to predict which twists and turns the process of negotiations will take, but at least the process of negotiations has resumed. The steps that I outline above are ones that I believe will enhance the security of South Korea and the United States and should be acceptable to the other parties. It will be up to the negotiators to determine what it takes to address Pyongyang's concerns: food aid alone will not suffice. Washington will have to address Pyongyang's fundamental insecurity and future energy and economic needs. Washington and Seoul will have to overcome domestic politics to do so. The young North Korean leader will have to be willing to seek a better future for his destitute people.

In the Long Term, Will Pyongyang Give Up Its Nuclear Weapons?

Although Pyongyang continues to claim that it is committed to the September 19 Joint Statement on the denuclearization of the Korean peninsula, it has made it clear that it is not prepared to give up its nuclear weapons any time soon. Nuclear weapons have become central to Pyongyang's projection of military might. North Korea views them as crucial to the regime's survival. They also appear to play a supportive role domestically for the regime and provide it with international diplomatic leverage.[37] The reasons why the regime

---

[37] Hecker, "Lessons Learned," pp. 44–56.

believes it needs the bomb are deeply rooted in history and, hence, are unlikely to be resolved by alliances with its neighbors, each of which, North Korea believes, has ulterior motivations. With its great mistrust of the United States, the regime will require much more than another security guarantee from Washington to make it feel secure.

Even if North Korea's security fears are assuaged, domestic factors may favor keeping the bomb. The external threat environment is used by the regime to justify the need for the bomb and the sacrifices that North Korea continues to ask its people to make. That threat also helps keep its people submissive and isolated from the international community. It also helps the regime control all information and to blind its people to progress in the rest of the world, especially south of the demilitarized zone. However, the recent exponential growth of cell-phone ownership in North Korea, which now tops around one million, will eventually undermine the regime's rigid control.[38]

It is unlikely that North Korea can be forced to give up the bomb. Realistically, military options are off the table unless North Korea initiates a conflict. Additionally, sanctions are mostly ineffective without China's support, but China will not support sanctions that bring Pyongyang to its knees. It wants peace and stability on the Korean peninsula, which Beijing believes is best achieved if Pyongyang reforms its economy and opens up along the lines of what China did some thirty years ago.

Pyongyang will give up its nuclear weapons only if the benefits of giving them up outweigh the cost of keeping them. The cost of its nuclear program has been, and continues to be, enormous. The resulting isolation from international commerce along with its military-first policy and centrally controlled economic system has led it

---

38 Alexandre Mansourov, "North Korea on the Cusp of Digital Transformation." Nautilus Institute Special Report, October 20, 2011.

to a state of abject poverty in contrast to its free-market southern neighbor. Ironically, Kim Jong-un may be trying to protect his regime against imagined external enemies while its primary threats are internal and economic—a situation perhaps not unlike that in the dying days of the Soviet Union.

North Korea's nuclear choice and current economic status stand in stark contrast to that of South Korea. Although South Korea explored the development of nuclear weapons,[39] it abandoned its fledgling weapon program because of heavy pressure from Washington and strengthened US security guarantees. As part of South Korea's drive to become an international economic powerhouse, it began to build a robust nuclear-power program, initially with Western technology and assistance, but eventually developing an impressive indigenous capability. South Korea has developed its nuclear program in a transparent manner, with good cooperation with the IAEA.[40]

Today, South Korea has 23 modern LWRs, producing 31 percent of its electricity.[41] It has a strong nuclear-research establishment in the Korean Atomic Energy Research Institute (KAERI). Together with

---

[39] South Korea first explored the plutonium route by conducting experiments to irradiate uranium rods in a research reactor and reprocessing them in a hot-cell laboratory in 1981. This was likely a precursor to designing a reprocessing plant, after US pressure stopped an attempted acquisition of such a plant from France. These experiments were disclosed and that activity was stopped. Past activities thus imply that the South Koreans were interested in obtaining fissile-material-separation capability, through either the uranium or the plutonium route. See, for instance NTI: South Korea Profile"; http://www.nti.org/e_research/profiles/SKorea/Nuclear/index.html.

[40] In 2004, South Korean scientists from the Korean Atomic Energy Research Institute (KAERI), however, conducted experiments in laser-isotopic enrichment of various elements, including uranium. These experiments were conducted without informing Korean Ministry officials or IAEA safeguards inspectors. This activity was stopped and fully reported to the IAEA in 2005.

[41] The World Nuclear Association states that South Korea plans to provide 59 perecnt of electricity from 40 nuclear plants by 2030; http://world-nuclear.org/info/inf81.html.

industrial giant Daewoo, KAERI recently signed a contract to build a research reactor for Jordan. South Korea now harbors realizable ambitions of becoming one of the world's leading exporters of nuclear-power plants. The December 2009 award of a $20.4 billion contract to build the first power plants in the United Arab Emirates stands as an example of this budding global role.[42] Now, South Korea has too much to lose economically to consider developing a nuclear-weapons option. In fact, it tries to be especially transparent and compliant so as not to jeopardize its export business. By contrast, the North has no nuclear electricity today and little prospect for indigenously developing any significant amounts for the next decade. North Korea's scientists and engineers are cut off from the rest of the world. Their exports have all been illicit and are drying up. They have a handful of bombs, but at a large price.

I draw the contrast between the North and South not to suggest that North Korea could have done as well if it had pursued nuclear electricity instead of bombs, but to demonstrate that North Korea could have much to gain in the future by trading its military program for a civilian one. Given its sorry economic state, North Korea has few other opportunities than to give up its nuclear-weapons program to improve its dire economic condition.

I remain optimistic in the long term. To stay in power, Kim Jong-un will have to improve the livelihood of his people. He will have to open up his country and reform the North's economy. If his security concerns can be assuaged by China, the United States, and South Korea, he may find that the nuclear card gives him significant bargaining leverage. Paradoxically, compared with a more democratic

---

[42] Margaret Coker, "U.A.E. Awards $20.4 billion Contract for Four Reactors in Breakthrough on World Stage for Korea Electric Power," *Wall Street Journal*, December 28, 2009; http://online.wsj.com/article/SB10001424052748704905704574621653002992302.html.

country, an autocracy like North Korea may find it easier to give up its weapons if doing so is seen to help the regime survive, because it does not have to deal with domestic opposition. The challenge for Washington and Seoul will be to focus on how to improve regional security while dealing with their own contentious domestic politics to encourage the North's transition.

Figure 1. **A time sequence of overhead images of the light-water reactor site tracking its development from September 2010 to February 2012. Credit: DigitalGlobe and GeoEye**

Figure 2. The experimental light-water reactor as of February 3, 2012. Credit: DigitalGlobe

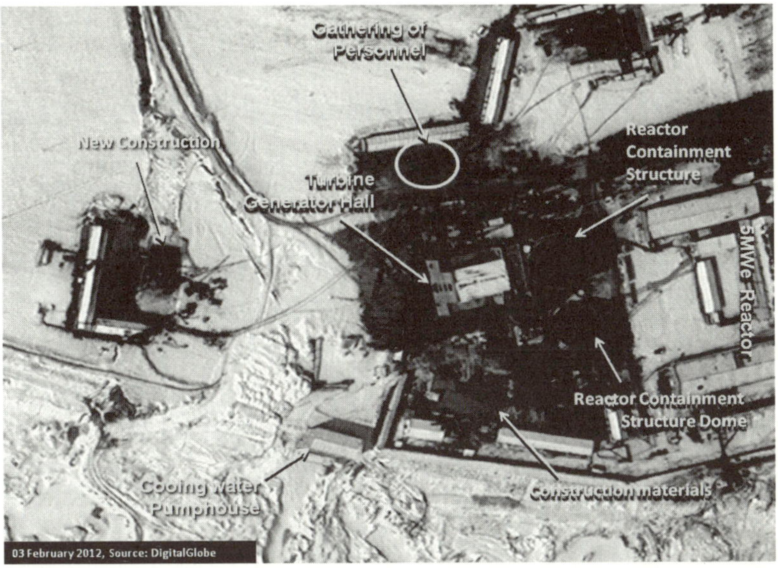

Figure 3. Three-dimensional model of the light-water reactor based on satellite images

Figure 4. A rough schematic of the floor plan of the cascade hall at the uranium-enrichment centrifuge facility (Building 4) in Yongbyon, November 12, 2010

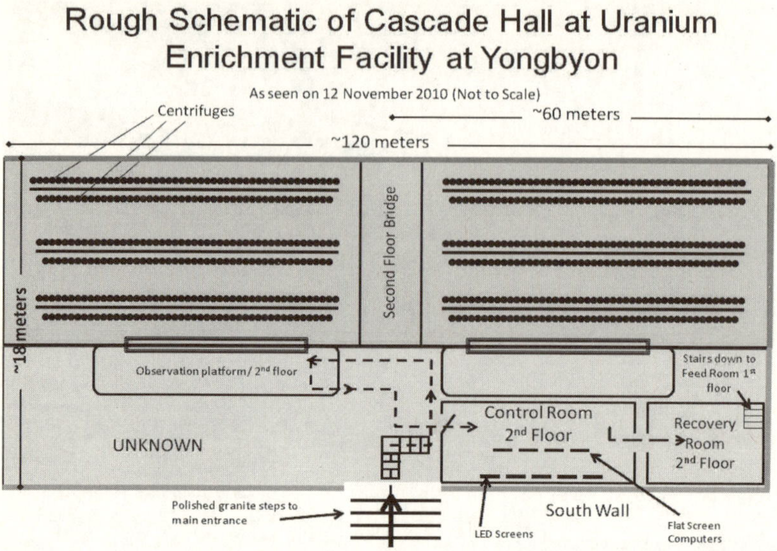

Figure 5. A three-dimensional model of Building 4 (the new uranium-enrichment centrifuge plant) in the fuel-fabrication plant, created using satellite images

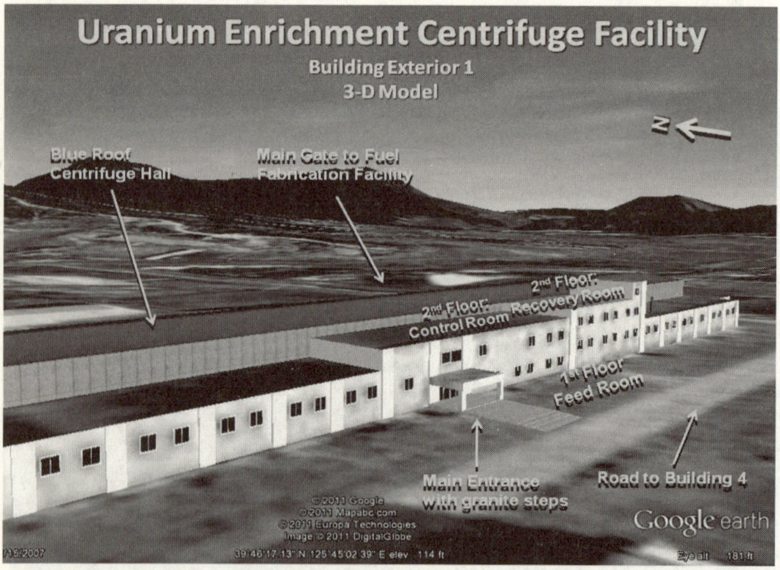

Chapter 6

Nuclear Power in China, Japan, and Korea: A Current and Future Assessment

Jian Cai

Nuclear technology uses the energy released by splitting the atoms of certain elements. It was first developed in the 1940s, and during the World War II research initially focused on producing bombs by splitting the atoms of either uranium or plutonium. But in the 1950s, attention turned to the peaceful purposes of nuclear fission; notably, for power generation. Today, only 8 countries are known to have a nuclear-weapons capability, but 30 countries use nuclear fission for

Figure 6. **Nuclear Electricity Production and Share of Total Electricity Production**

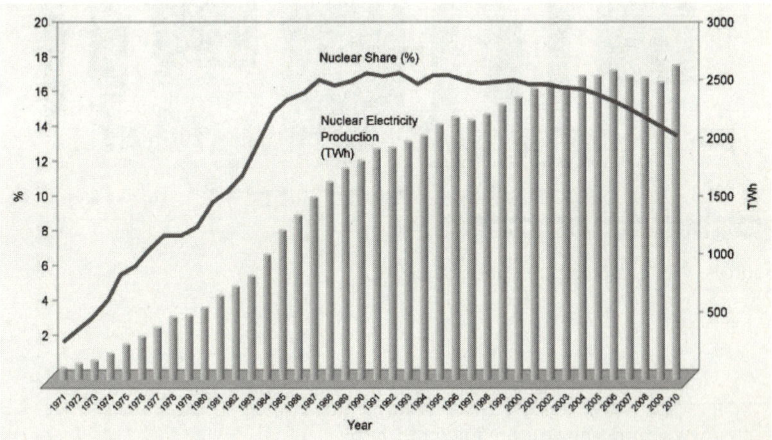

energy purpose. These 30 countries operate some 440 commercial nuclear-power reactors, with a total installed capacity of more than 377,000 MWe, which supplies almost 14 percent of global electricity needs.[1]

Sixteen of the the 30 countries depend on nuclear power for at least one-quarter of their electricity. France ranks first with around 75 percent of its power from nuclear energy, while Belgium, Bulgaria, the Czech Republic, Hungary, Slovakia, South Korea, Sweden, Switzerland, Slovenia, and Ukraine get 30 percent or more. Japan, Germany, and Finland get more than 15 percent of their power from nuclear energy; the United States gets 20 percent. But China gets only 1.9 percent of its power from nuclear energy.

Figure 7. **Fuel for Electricity Generation 2008**

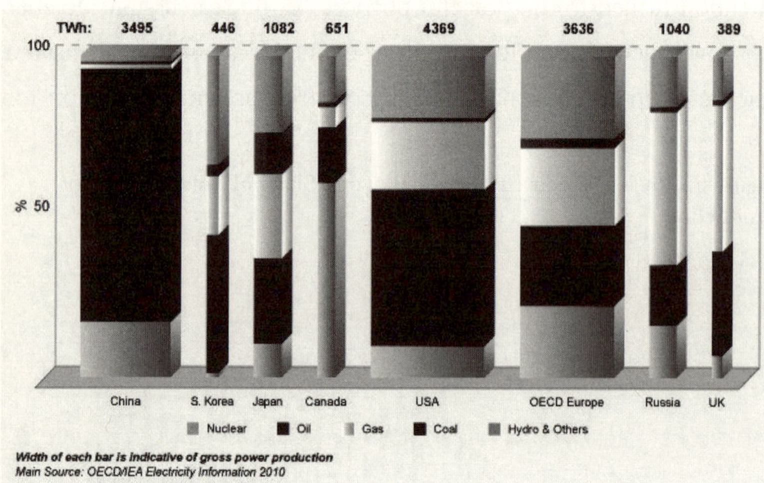

Sources: WNA, data to publication date. IAEA.

---

[1] http://www.world-nuclear.org/info/inf01.html.

Notwithstanding these numbers, if we examine the international nuclear industry carefully, we find that it cannot stop a slow decline of nuclear energy. Compared with previous years, not enough new units come online. Up to April 1, 2011, there were 437 nuclear reactors operating in the world—seven fewer than in 2002. According to International Atomic Energy Agency (IAEA) statistics, 64 reactors are "under construction" in 14 countries. By comparison, at the peak of the industry's growth phase in 1979, 233 reactors were being built concurrently. In 2008, for the first time since the beginning of the nuclear age, no new unit was started up. Two were added in 2009, five in 2010, and two in the first three months of 2011. By contrast, 11 reactors were shut down during the same time period.[2]

Besides this, the world's reactor fleet is aging quickly. The average age of the world's operating nuclear power plants is 26 years. Considering that the average age of the 130 units that already have been closed is about 22 years and the lessons from the Fukushima nuclear-power plants (operating around 40 years), the operating age of the units will be looked at in a quite different manner; for example, the German government decided to suspend operation of all reactors over 30 years old immediately. If we assumes an average lifetime of 40 years for all operating and inconstruction reactors, we can easily estimate how many plants would be shut down year by year and how many plants should have to come on line to maintain the same number of operating plants. Even if the installed capacity level could be maintained, the number of operating reactors will decline over the coming years unless lifetime extensions beyond 40 years become the widespread standard.

Moreover, it is now evident that nuclear-power development

---

[2] Mycle Schneider: "The World Nuclear Industry Status Report, 2010–2011," Worldwatch Institute, Washington, DC.

cannot keep up with the pace of its renewable energy competitors. In 2010, for the first time, worldwide cumulated installed capacity of wind turbines (193 gigawatts), biomass and waste-to-energy plants (65 GW), and solar power (43 GW) reached 381 GW, outpacing the installed nuclear capacity of 375GW.

In 2009, nuclear-power plants generated 2,558 TWh of electricity, about 2 percent less than the previous year—just as the World Nuclear Association said, it was "another drop in nuclear generation," for the fourth year in a row. In 2010, 16 of the 30 countries operating nuclear-power plants maintained their nuclear share in electricity generation, while nine decreased their share and five increased their share.

The three factors described above point to the role of nuclear power declining steadily. But is such a development in fact what is truly happening? According to the International Atomic Energy Agency (IAEA), more and more countries want to introduce nuclear power: "Some 60 countries have turned to the IAEA for guidance," and now there are more units under construction worldwide. Does this mean a "nuclear renaissance"? Or is the phenomenon limited to only some countries, with China alone counting for 60 percent of the new projects?

Nuclear Power in China

The Reason for China's Nuclear-Power Development

When China adopted its open-door and reforming policy in 1978, the economy began to develop rapidly. Rapid growth in demand has given rise to power shortages. Electricity consumption in 2011 rose 11.7 percent to 4,693 billion kWh, according to the National Energy

Bureau. In 2010, it increased 14.56 percent to 4,190 billion kWh, according to the China Electricity Council, corresponding with a 10 percent growth in GDP. Some 3,090 billion kWh of this was in industry. Most of China's electricity is produced from fossil fuels (80 percent from coal, 2 percent from oil, 1 percent from gas in 2006) and hydropower (15 percent). Since the coal is the main energy source and most reserves are in the north or northwest, there is an enormous logistical problem. Nearly one-half the country's rail capacity is used in transporting coal. The US Energy Information Administration predicts that China's coal consumption is expected to exceed 4 billion tonnes per year by 2015—one-half of the world total. Because of the heavy reliance on old, coal-fired plants, electricity generation accounts for much of the country's air pollution, China recently overtook the United States as the world's largest contributor to carbon dioxide emissions. The economic loss due to pollution is put by the World Bank at almost 6 percent of GDP.[3] Due to increasing concerns about air quality, climate change, and fossil-fuel shortages, nuclear power has been looked to as an alternative to coal power in China.[4]

China's Ambition for Nuclear Power

China came relatively late to the civil nuclear industry, starting construction of its first commercial reactor only in 1985. As of February 2012, China had 16 reactors in operation, which in 2009 provided only 1.9 percent of the country's electricity, the lowest nuclear share of any country. Despite, or maybe because of, its late arrival to the nuclear field, China now has an impressive recent history of construction starts. As of February 2012, there were 26 reactors under construction.

---

[3] IAEA, "Power Reactor Information System," 2011; http://www.iaea.org/programmes/a2/.

[4] World Nuclear Association, "Nuclear Power in China," July 2, 2010; http://www.world-nuclear.org/info/inf63.html, retrieved March 2, 2012.

China's concerted nuclear expansion began with the National Development and Reform Commission's (NDRC's) Tenth Economic Plan for the years 2001 to 2005. It set a big goal for Chinese nuclear power: during the tenth five-year plan (2001–2005), the installed capacity of new starting nuclear power should get to 6,000 megawatts (mw) and make the total capacity of operating nuclear power units to 15,000 megawatts (mw) in 2010. The Tenth Five-Year Plan (2001–05) incorporated the construction of four projects, which consist of eight nuclear reactors. The four units in the Tenth Plan were Ling Ao Phase II (also known as Ling Dong) in Guangdong Province and Qinshan Phase II, in Zhejiang Province, plus two others that, bwhen the timeline for contracts was extended, were put into the 11th plan: Sanmen, in Zhejiang Province, and Yangjiang, in Guangdong Province.

The 11th Economic Plan, for 2006 to 2010, set even more ambitious goals than the 10th for new nuclear-plant construction and marked a watershed in China's commitment to third-generation reactors, such as the Sanmen plant in Zhejiang and the Haiyang plant in Shandong Province. The 11th Five Year Plan (2006–10) had firmer environmental goals than previously, including reduction of 20 percent in the amount of energy required per unit of GDP (a 4 percent reduction per year). As well as the Sanmen and Yangjiang projects that were slipped from the 10th Plan, nuclear-power developments originally proposed in the 11th Plan included reactors at Hongyanhe (four), Haiyang (two), Fuqing (two), and Taishan (two), all of which are now under construction. Two at Lufeng and two at Hongshiding are delayed.

The 12th Five-Year Plan (2011–15) includes construction starts on phase II of Tianwan, Hongyanhe, Sanmen, and Haiyang, as well as phase I of inland sites Taohuajiang, Xianning, and Pengze (two reactors each, except Taohuajiang, which will have four). By the end of the 12th Five-Year Plan, some 25 GWe of new capacity is planned to be operational, making some 40 GWe, and 45 GWe more may be added by the end of the 13th Five-Year Plan. However, government targets for nuclear power have been increasing. As of June 2010, the official projections for installed nuclear capacity were 70–80 GWe by 2020, 200 GWe by 2030 and 400–500 GWe by 2050. Following the Fukushima accident and the consequent pause in approvals for new plants, the target is now more like 60–70 GWe by 2020.

In December 2011, the National Energy Administration (NEA) said that China will make nuclear energy the foundation of its power-generation system in the next "10 to 20 years," adding as much as 300 GWe of nuclear capacity over that period. Two weeks earlier, the NDRC (National Development and Reform Commision) vice-director said that China would not swerve from its goal of greater reliance on nuclear power. The former head of the NEA said that full-scale construction of nuclear plants would resume in March 2012. More than 16 provinces, regions, and municipalities have announced intentions to build nuclear-power plants in the 12th Five-Year Plan (2011–15).

Power Generated by Nuclear Plants in China

Nuclear power has an important role in China's economic growth, especially in the coastal areas, remote from the coalfields and where the economy is developing rapidly. Generally, nuclear plants can be built close to centers of demand, whereas suitable wind and hydro sites are remote from demand. Moves to build nuclear power commenced in 1970, and about 2005 the industry moved into a rapid-development phase. In China, most nuclear-power plants are located on the coast and generally use seawater for cooling a direct, once-through cycle. But as the different technology has been drawn from France, Canada, Russia, and the United States, the new technology seems suitable to be installed inland.

China used French, Russian, and Canadian designs to kick-start its commercial nuclear-power program. The first commercial reactors were built at Daya Bay, in Guangdong Province. The reactors are 994 MW, three-loop, French, pressurized-water-reactor (PWR) units supplied by Framatome (now Areva), with GEC-Alstom—supplied steam turbines. Construction started in August 1987, and commercial operation of the two units commenced in February and May 1994. At Ling Ao, which adjoins Daya Bay, two more reactors were constructed using the same technology, but with 30 percent localization. The Ling Ao reactors started generating commercially in 2002 and 2003. The Tianwan Phase I at Lianyungang in Jiangsu Province is a Russian-designed AES-91 power plant consisting of two 1060 MW VVER reactors. The first unit entered commercial operation in June 2007, with the second following it on line in August 2007.

China's first "home-grown" nuclear-power plant was a 300 MW PWR (CNP-300) built at Qinshan, 100 km southwest of Shanghai. The Shanghai Nuclear Engineering Research and Design Institute (SNERDI) was responsible for the design. Another SNERDI design, a 600 MW PWR (CNP-600), was built at Qinshan for Phase II. The first unit entered commercial operation in April 2002. The second came on in May 2004.

China looked to Canada for reactor designs for Qinshan Phase III. This consists of two 665 MW Candu 6 pressurized heavy-water reactors (PHWR), designed by Atomic Energy of Canada (AECL). Following State Council approval in September 2004 for the construction of Generation III reactors at Sanmen and Yangjiang, an open bidding process was started. Reactor vendors Westinghouse (AP1000), Areva (EPR), and Atomstroyexport (VVER-1000 model V-392) bid for the contracts. Willingness to transfer the technology was a key priority in assessing the bids, as the Chinese strategy is to move swiftly toward domestic production of the new reactors. The Westinghouse AP1000 came out of the bidding process as the clear winner, and at the end of 2006 a decision was taken to build four units of the Westinghouse AP1000 at Sanmen (in Guandong) and Haiyang (in Shandong). More than 50 of China's proposed new nuclear reactors look likely to be AP1000s (or CAP1000s, as they are likely to be known).

The AP1000 technology transfer is going on in parallel with current construction at Sanmen and Haiyang. Westinghouse is transferring around 60,000 documents and providing training classes to a range of Chinese scientists and technicians. It also established a job-shadowing program in which 120 core operating staff at the four stations spend two months doing internships at operating US nuclear-power plants.

Although Areva did not win the bid for the first four Generation III units, in 2007 it signed contracts for two 1650 MW EPRs (Evolutionary Power Reactors) to be built at Taishan on China's southeast coast.

Figure 8. **China NPPs Distribution**

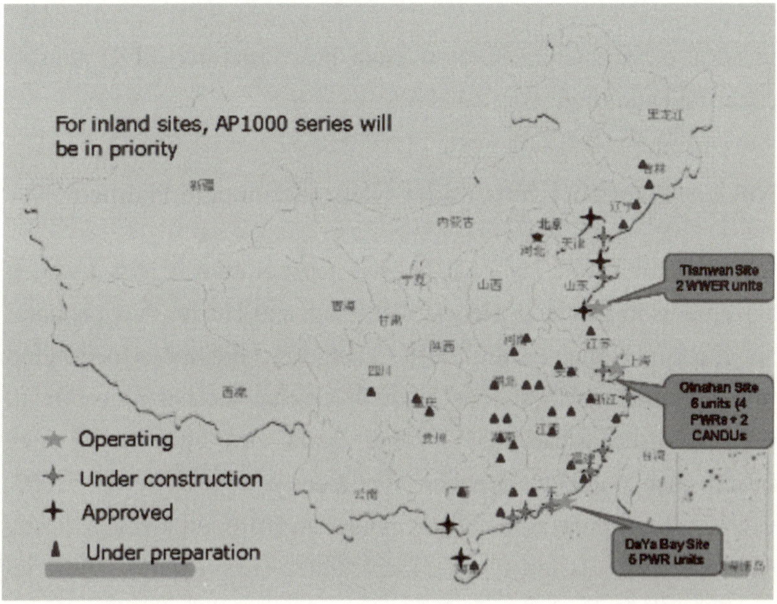

Table 2. **Operating Nuclear Reactors in China**

Units	Province	Net capacity (each)	Type	Operator	Commercial operation
Daya Bay 1&2	Guangdong	944 MWe	PWR (French M310)	CGNPC	1994
Qinshan Phase I	Zhejiang	279 MWe	PWR (CNP-300)	CNNC	April 1994
Qinshan Phase II, 1-4	Zhejiang	610 MWe	PWR (CNP-600)	CNNC	2002, 2004, 2010, (2012)
Qinshan Phase III, 1&2	Zhejiang	665 MWe	PHWR (Candu 6)	CNNC	2002, 2003
Ling Ao Phase I, 1&2	Guangdong	935 MWe	PWR (French M310)	CGNPC	2002, 2003
Tianwan 1&2	Jiangsu	1000 MWe (VVER-1000)	PWR	CNNC	2007, 2007
Ling Ao Phase II, 1&2	Guangdong	1037 MWe	PWR (CPR-1000)	CGNPC	Sept 2010, Aug 2011
Total: 15		11,881 MWe			

In addition, the China Experimental Fast Reactor (CEFR) is grid-connected and producing 20 MWe.

Nuclear Plants in China under Construction and Planned

In 2006, the China National Nuclear Corporation (CNNC) signed agreements in Liaoning, Hebei, Shandong, and Hunan Provinces and six cities in Hunan, Anhui, and Guangdong Provinces to develop nuclear projects. CNNC has pointed out that there is room for 30 GWe of further capacity by 2020 in coastal areas, and maybe more inland, such as in Hunan, "where conditions permit" In October 2007, CNNC's list of projects included Chuanshan (Jiangsu Province), Jiyang (Anhui), Hebao Island (Guangdong), Shizu (Chongqing), Xudabao/

Xudapu (Liaoning), and Qiaofushan (Hebei), among others.

In November 2007, the NDRC said that the government had budgeted CNY 450 billion ($65 billion) to build nuclear-power capacity by 2020. It had selected 13 coastal sites to accommodate 59.46 GWe.

In mid-2009, the Huaneng Nuclear Power Development Co.—a subsidiary of the China Huaneng Group (CHNG)—opened an office in the city of Yingtan in China's inland Jiangxi Province for the development of a new nuclear-power plant in the area. This is one of five sites for nuclear plants after Rongcheng, which CHNG was reported in May to have selected—Cangnan in Zhejiang Province, Huaining in Anhui, Xuyi in Jiangsu, and Xiapu in Fujian being the others.

In December 2009, CGNPC signed a CNY 5.3 billion ($776 million) nuclear island installation contract with the China Nuclear Power Engineering Group Co. (CNPEC), apparently covering Ningde 3 and 4 in Fujian, Yangjiang 3 and 4 in Guangdong, Fangchenggang 1 and 2 in Guangxi, and Taishan 1 and 2 in Guangdong. This is the largest contract of its kind in China.

***Table** 3.* **Nuclear Reactors under Construction and Planned in China**

Plant	Province	MWe gross	Reactor model	Project control	Construction start	Operation
Hongyanhe units 1-4	Liaoning Guangdong	4x1080	CPR-1000	CGNPC, with CPI	**8/07, 4/08, 3/09, 8/09**	10/12, 2013, 2014
Ningde units 1-4	Fujian	4x1080	CPR-1000	CGNPC, with Datang	**2/08, 11/08, 1/10, 9/10**	12/12, 2013, 2014, 2015
Fuqing units 1&2	Fujian	2x1080	CPR-1000	CNNC	**11/08, 6/09**	11/13, 9/14
Yangjiang units 1-4	Guangdong	4x1080	CPR-1000	CGNPC	**12/08, 8/09, 11/10,** 3/11*	8/13, 2014, 2015, 2016

Table 3. **Nuclear Reactors under Construction and Planned in China (Continued)**

Plant	Province	MWe gross	Reactor model	Project control	Construction start	Operation
Fangjiashan units 1&2	Zhejiang	2x1080	CPR-1000	CNNC	**12/08, 7/09**	12/13, 10/14
Sanmen units 1&2	Zhejiang	2x1250	AP1000	CNNC	**3/09, 12/09**	11/13, 9/14
Haiyang units 1&2	Shandong	2x1250	AP1000	CPI	**9/09, 6/10**	5/14, 3/15
Taishan units 1&2	Guangdong	2x1770	EPR	CGNPC	**10/09, 4/10**	10/13, 11/14
Hongyanhe units 5&6	Liaoning	2x1080	CPR-1000	CGNPC, with CPI	2011, 2011?	2015
Shandong Shidaowan	Shandong	210	HTR-PM	Huaneng	**mid 2011**	2015
Fangchenggang units 1&2	Guangxi	2x1080	CPR-1000	CGNPC	**7/10,** 2011	2015, 2016
Fuqing units 3&4	Fujian	2x1080	CPR-1000	CNNC	**7/10,** 2011*	7/15, 5/16
Fuqing units 5&6	Fujian	2x1080	CPR-1000 or CNP1000	CNNC	2011, 2011*	–
Changjiang units 1&2	Hainan	2x650	CNP-600	CNNC & Huaneng	**4/10, 11/10**	2014, 2015
Hongshiding (Rushan) units 1&2	Shandong	2x1080	CPR-1000	CNEC/ CNNC	Deferred from 2009?	2015
Yangjiang units 5&6	Guangdong	2x1080	CPR-1000	CGNPC	2011*	2017
Ningde units 5&6	Fujian	2x1080	CPR-1000	CGNPC		
Xianning (Dafan) units 1&2	Hubei	2x1250	AP1000	CGNPC	2011 or 2015	2015?

Table 3. **Nuclear Reactors under Construction and Planned in China (Continued)**

Plant	Province	MWe gross	Reactor model	Project control	Construction start	Operation
Taohuajiang units 1-4	Hunan	4x1250	AP1000	CNNC	2011	4/2015-2018?
Pengze units 1&2	Jiangxi	2x1250	AP1000	CPI	2011 or 2015	2015?
Xudabao / Xudapu units 1&2	Liaoning	2x1250	AP1000	CPI, with Datang	9/11, ? delayed	
Sanmen units 3&4	Zhejiang	2x1250	AP1000	CNNC		
Haiyang units 3&4	Shandong	2x1250	AP1000	CPI	2010?	
Xiaomoshan units 1&2	Hunan	2x1250	AP1000	CPI	2012	
Longyou (Zhexi) units 1&2	Zhejiang	2x1250	AP1000	CNNC	2012	
Sanming units 1&2	Fujian	2x880	BN-800	CNNC	2013	2019, 20
Zhangzhou units 1&2	Fujian	2x100?	SMR: ACP100?	CNNC & Guodian	2011	
Yanjiashan/ Wanan/Ji'an	Jiangxi	2x1250	AP1000	CNNC	2012	
Shaoguan units 1-4	Guangdong (inland)	4x1250	AP1000	CGNPC	2013	
Tianwan units 5&6	Jiangsu	2x1080	CPR-1000	CNNC	12/12, 8/13	
Tianwan units 3&4	Jiangsu	2x1060	VVER-1000	CNNC	?, ?	
Wuhu units 1&2	Anhui	2x1250	AP1000	CGNPC	2012	8/2016

Table 3. **Nuclear Reactors under Construction and Planned in China (Continued)**

Plant	Province	MWe gross	Reactor model	Project control	Construction start	Operation
Lianyungang units 1&2	Jiangsu	2x1080	CPR-1000	CGNPC		
Shanwei (Lufeng) units 1&2	Guangdong	2x1080	CPR-1000	CGNPC	late 2012?	
⋮	⋮	⋮	⋮			
Total: 77		85,750 MWe				

Where construction has started, the dates are marked in bold. Those here not under construction are marked as 'planned' in the WNA reactor table. At December 1, 2011, 26 were under construction: 27,640 MWe; 51 were planned: 57,480 MWe (gross). Fangjiashan is sometimes shown as a development of Qinshan Phase I. * approved, but construction delayed post-Fukushima.

Nuclear Power in Japan

Japan's Nuclear Policy

People always wonder that Japan is the only country in the world that has suffered the devastating effects of nuclear weapons in history, but nowadays Japan has embraced the nuclear technology eagerly and depended on nuclear power deeply. According to IAEA statistics, 50 reactors are operating now and another two reactors are under construction. These 50 reactors provide almost 30 percent of the country's total electricity production. There is a plan to increase this to 41 percent by 2017, and to 50 percent by 2030. So why did Japan want to develop the nuclear power so eagerly? The only reason for Japan to develop nuclear power is to sustain economic growth.

Japan has few natural resources of its own: it depends on imports for some 84 percent of its primary energy needs. Initially, it was dependent on fossil-fuel imports, particularly oil from the Middle East; for example, oil fuelled 66 percent of electricity in 1974. This geographical and commodity vulnerability constrained the economic development of Japan. When the oil shock hit in 1973, the situation became critical. The Japanese government reevaluated its energy policy and decided to reduce the country's dependence on oil imports, which eventually resulted in a major nuclear-construction program.[5]

The Kyoto Protocol is a protocol to the United Nations Framework Convention on Climate Change (UNFCCC or FCCC), aimed at fighting global warming, initially adopted in 1997.[6] This protocol affected the energy policies of many countries.

In March 2002, the Japanese government announced that Japan would rely heavily on nuclear energy in order to achieve greenhouse-gas emission-reduction goals set by the Kyoto Protocol.[7] A ten-year energy plan called for an increase in nuclear-power generation by about 30 percent by 2011. In June 2002, a new Energy Policy Law giving greater authority to the government in establishing the energy infrastructure for economic growth, a further move away from dependence on fossil fuels. In November 2002, the Japanese government announced that it would introduce a tax on coal for the first time, and at the same time the Ministry of Economy, Trade, and Industry (METI) would reduce its power-source development tax,[8] including that

---

5 World Nuclear Association, "Nuclear Power in Japan"; http://www.world-nuclear.org/info/default.aspx?id=344&terms=nuclear%20power%20in%20japan.

6 http://en.wikipedia.org/wiki/Kyoto_Protocol.

7 G-8 Governments—Nuclear Energy Policies; http://www.world-nuclear.org/reference/g8-policies.html.

8 Cooperative Research Center for Coal in Sustainable Development, Energy Scenarios—Japan; http://www.ccsd.biz/publications/files/TA/TA%2046%20Japan%20Energy%20scenarios.pdf.

applying to nuclear generation, by 15.7 percent. All these developments paved the way for an increased role for nuclear energy.

In 2004, Japan's Atomic Industrial Forum released a report on the future prospects for nuclear power in Japan. According to the report, the projected nuclear-generating capacity will get to 90 GWe in 2050, which means doubling both nuclear-generating capacity and nuclear share to about 60 percent of total power produced. In July 2005, the Atomic Energy Commission reaffirmed the above policy for nuclear power.

In 2008, METI initiated the Cool Earth 50 energy innovative-technology plan, which indicated that Japan should manage to gain the 54 percent reduction in CO2 emissions (from 2000 levels) by 2050 and 90 percent reduction by 2100. That means the nuclear energy would contribute about 60 percent of primary energy in 2100 (compared with 10 percent now).[9] Two years later, METI made a new plan aimed at increasing energy self-sufficiency to 70 percent by 2030, for both energy security and CO2 emission reduction. To gain this goal, nuclear power would play a big part in implementing the plan, and new reactors will be required as well as achieving a 90 percent capacity factor across all plants.

METI's 2010 electricity-supply plan also showed that Japan's nuclear capacity would grow by 12.94 GWe by 2019, and the share of supply would grow from 2007's 25.4 percent to 41 percent in 2019. But after the Fukushima accident, the Energy and Environment Council chaired by the minister for national policy decided to reduce future dependence on nuclear power in July, and in September Prime Minister Noda said he expected the country to reduce its dependency on nuclear power in the medium and long term.

---

[9] Ministry of Economy, Trade and Industry, Cool Earth 50 Energy Innovative Technology Program; http://www.iae.or.jp/research/project/Cool_Earth08_e/CoolEarth_RM.pdf.

Japan's Nuclear Programs

In 1954, Japan started its nuclear-research program and the following year introduced the Atomic Energy Basic Law, which strictly limits the use of nuclear technology to peaceful purposes. In 1956, the Atomic Energy Commission and other nuclear-energy-related organisations were established under this law, such as, among others, the Science and Technology Agency, the Japan Atomic Energy Research Institute (JAERI), and the Atomic Fuel Corporation (renamed PNC in 1967).

Japan's first reactor was a prototype boiling-water reactor, which ran from 1963 to 1976 and provided a large amount of information for later commercial reactors. Its first commercial nuclear-power reactor was one imported from the United Kingdom that had been operated from 1966 to 1998. In the 1970s, light-water reactors (LWRs), boiling-water reactors (BWRs), and pressurized-water reactors (PWRs) were constructed. At by the end of the 1970s, Japanese industry had largely established its own domestic nuclear-power production capacity.

Because of reliability problems with the earliest reactors, MITI and the nuclear-power industry launched the LWR Improvement and Standardization Program in 1975, which aimed to standardize LWR designs and to change the BWR and PWR into the Advanced BWR (ABWR) and the Advanced PWR (APWR) by 1985.

***Table 4.* Power Reactors Operating in Japan**

Name	Type	Status	Location	Reference Unit Power [MW]	Gross Electrical Capacity [MW]	First Grid Connection
FUGEN ATR	HWLWR	Permanent Shutdown	TSURUGA	148	165	1978-07-29
FUKUSHIMA-DAIICHI-1	BWR	Permanent Shutdown	OHKUMA-MACHI	439	460	1970-11-17

Table 4. **Power Reactors Operating in Japan (Continued)**

Name	Type	Status	Location	Reference Unit Power [MW]	Gross Electrical Capacity [MW]	First Grid Connection
FUKUSHIMA-DAIICHI-2	BWR	Permanent Shutdown	OHKUMA-MACHI	760	784	1973-12-24
FUKUSHIMA-DAIICHI-3	BWR	Permanent Shutdown	OHKUMA-MACHI	760	784	1974-10-26
FUKUSHIMA-DAIICHI-4	BWR	Permanent Shutdown	OHKUMA-MACHI	760	784	1978-02-24
FUKUSHIMA-DAIICHI-5	BWR	Operational	OHKUMA-MACHI	760	784	1977-09-22
FUKUSHIMA-DAIICHI-6	BWR	Operational	OHKUMA-MACHI	1067	1100	1979-05-04
FUKUSHIMA-DAINI-1	BWR	Operational	NARAHA-MACHI	1067	1100	1981-07-31
FUKUSHIMA-DAINI-2	BWR	Operational	NARAHA-MACHI	1067	1100	1983-06-23
FUKUSHIMA-DAINI-3	BWR	Operational	NARAHA-MACHI	1067	1100	1984-12-14
FUKUSHIMA-DAINI-4	BWR	Operational	NARAHA-MACHI	1067	1100	1986-12-17
GENKAI-1	PWR	Operational	GENKAI-CHO	529	559	1975-02-14
GENKAI-2	PWR	Operational	GENKAI-CHO	529	559	1980-06-03
GENKAI-3	PWR	Operational	GENKAI-CHO	1127	1180	1993-06-15
GENKAI-4	PWR	Operational	GENKAI-CHO	1127	1180	1996-11-12
HAMAOKA-1	BWR	Permanent Shutdown	OMAEZAKI-SHI	515	540	1974-08-13
HAMAOKA-2	BWR	Permanent Shutdown	OMAEZAKI-SHI	806	840	1978-05-04
HAMAOKA-3	BWR	Operational	OMAEZAKI-SHI	1056	1100	1987-01-20

Table 4. **Power Reactors Operating in Japan (Continued)**

Name	Type	Status	Location	Reference Unit Power [MW]	Gross Electrical Capacity [MW]	First Grid Connection
HAMAOKA-4	BWR	Operational	OMAEZAKI-SHI	1092	1137	1993-01-27
HAMAOKA-5	BWR	Operational	OMAEZAKI-SHI	1325	1380	2004-04-26
HIGASHI DORI 1 (TOHOKU)	BWR	Operational	Higashidori-mura	1067	1100	2005-03-09
IKATA-1	PWR	Operational	IKATA-CHO	538	566	1977-02-17
IKATA-2	PWR	Operational	IKATA-CHO	538	566	1981-08-19
IKATA-3	PWR	Operational	IKATA-CHO	846	890	1994-03-29
JPDR	BWR	Permanent Shutdown	TOKAIMURA	12	13	1963-10-26
KASHIWAZAKI KARIWA-1	BWR	Operational	KASHIWAZAKI-SHI	1067	1100	1985-02-13
KASHIWAZAKI KARIWA-2	BWR	Operational	KASHIWAZAKI-SHI	1067	1100	1990-02-08
KASHIWAZAKI KARIWA-3	BWR	Operational	KASHIWAZAKI-SHI	1067	1100	1992-12-08
KASHIWAZAKI KARIWA-4	BWR	Operational	KASHIWAZAKI-SHI	1067	1100	1993-12-21
KASHIWAZAKI KARIWA-5	BWR	Operational	KASHIWAZAKI-SHI	1067	1100	1989-09-12
KASHIWAZAKI KARIWA-6	BWR	Operational	KASHIWAZAKI-SHI	1315	1356	1996-01-29
KASHIWAZAKI KARIWA-7	BWR	Operational	KASHIWAZAKI-SHI	1315	1356	1996-12-17
MIHAMA-1	PWR	Operational	MIHAMA	320	340	1970-08-08
MIHAMA-2	PWR	Operational	MIHAMA	470	500	1972-04-21
MIHAMA-3	PWR	Operational	MIHAMA	780	826	1976-02-19

Table 4. **Power Reactors Operating in Japan (Continued)**

Name	Type	Status	Location	Reference Unit Power [MW]	Gross Electrical Capacity [MW]	First Grid Connection
MONJU	FBR	Long-term Shutdown	TSURUGA	246	280	1995-08-29
OHI-1	PWR	Operational	OHI	1120	1175	1977-12-23
OHI-2	PWR	Operational	OHI	1120	1175	1978-10-11
OHI-3	PWR	Operational	OHI	1127	1180	1991-06-07
OHI-4	PWR	Operational	OHI	1127	1180	1992-06-19
OHMA	BWR	Under Construction	OHMA	1325	1383	2199-01-01
ONAGAWA-1	BWR	Operational	ONAGAWA, ISHINOMAKI	498	524	1983-11-18
ONAGAWA-2	BWR	Operational	ONAGAWA, ISHINOMAKI	796	825	1994-12-23
ONAGAWA-3	BWR	Operational	ONAGAWA, ISHINOMAKI	796	825	2001-05-30
SENDAI-1	PWR	Operational	SATSUMASENDAI	846	890	1983-09-16
SENDAI-2	PWR	Operational	SATSUMASENDAI	846	890	1985-04-05
SHIKA-1	BWR	Operational	SHIKA-MACHI	505	540	1993-01-12
SHIKA-2	BWR	Operational	SHIKA-MACHI	1108	1206	2005-07-04
SHIMANE-1	BWR	Operational	MATSUE	439	460	1973-12-02
SHIMANE-2	BWR	Operational	MATSUE	789	820	1988-07-11
SHIMANE-3	BWR	Under Construction	MATSUE	1325	1373	2199-01-01
TAKAHAMA-1	PWR	Operational	TAKAHAMA	780	826	1974-03-27
TAKAHAMA-2	PWR	Operational	TAKAHAMA	780	826	1975-01-17
TAKAHAMA-3	PWR	Operational	TAKAHAMA	830	870	1984-05-09

Table 4. **Power Reactors Operating in Japan (Continued)**

Name	Type	Status	Location	Reference Unit Power [MW]	Gross Electrical Capacity [MW]	First Grid Connection
TAKAHAMA-4	PWR	Operational	TAKAHAMA	830	870	1984-11-01
TOKAI-1	GCR	Permanent Shutdown	TOKAI MURA	137	166	1965-11-10
TOKAI-2	BWR	Operational	TOKAI MURA	1060	1100	1978-03-13
TOMARI-1	PWR	Operational	TOMARI VILLAGE	550	579	1988-12-06
TOMARI-2	PWR	Operational	TOMARI VILLAGE	550	579	1990-08-27
TOMARI-3	PWR	Operational	TOMARI VILLAGE	866	912	2009-03-20
TSURUGA-1	BWR	Operational	TSURUGA CITY	340	357	1969-11-16
TSURUGA-2	PWR	Operational	TSURUGA CITY	1108	1160	1986-06-19

*The above data are from the PRIS database. updated 03 05, 2012.[10]

Table 5. **Reactors under Construction in Japan**

Reactor	Type	Gross capacity	Utility	Construction start	Operation
Shimane 3	ABWR	1373 MWe	Chugoku	December 2005, suspended 2011	3/2012 but deferred
Ohma 1	ABWR	1383 MWe	EPDC/ J-Power	May 2010, suspended 3/11	11/2014 confirmed Dec 2011
Total (2)		2756 MWe			

---

10 http://pris.iaea.org/public/CountryStatistics/CountryDetails.aspx?current=JP.

Table 6. **Reactors Planned and Proposed in Japan**

Reactor	Type	MWe gross (each)	Utility	Start* construction	Start* operation
Tsuruga 3	APWR	1538	JAPC	3/2012	7/2017
Tsuruga 4	APWR	1538	JAPC	3/2012	7/2018
Higashidori 1 Tepco	ABWR	1385	Tepco	4/2011 (deferred)	3/2017
Kaminoseki 1	ABWR	1373	Chugoku	6/2012 (deferred 3/11)	3/2018
Sendai 3	APWR	1590	Kyushu	3/2014 (deferred 4/11)	12/2019
Higashidori 2 Tepco	ABWR	1385	Tepco	2014?	2019 or later
Hamaoka 6	ABWR	1380	Chubu	2016 or later	2020 or later
Higashidori 2 Tohoku	ABWR	1385	Tohoku	2016	2021 or later
Namie-odaka	BWR?	825?	Tohoku	2017	3/2021
Kaminoseki 2	ABWR	1373	Chugoku	2018 (deferred 6/11)	2022
Total Planned (10)		13,772 MWe			
Fukushima I-7	ABWR	1380	Tepco	4/2012 (suspended)	
Fukushima I-8	ABWR	1380	Tepco	4/2012 (suspended)	
not identified	ABWR?	4000 total	Chubu		2030
Total proposed (5)		6760 MWe			

* According to the METI FY2010 plan, unless updated by the company. The Tsuruga 3–4 schedule has slipped by 16 months. Tsuruga 3–4 and Tepco's Higashidori 1 are undergoing final safety assessment by regulatory authorities. The units listed as Fukushima I–7 and 8 are likely to be built elsewhere.

Figure 9. Japan NPPs Distribution

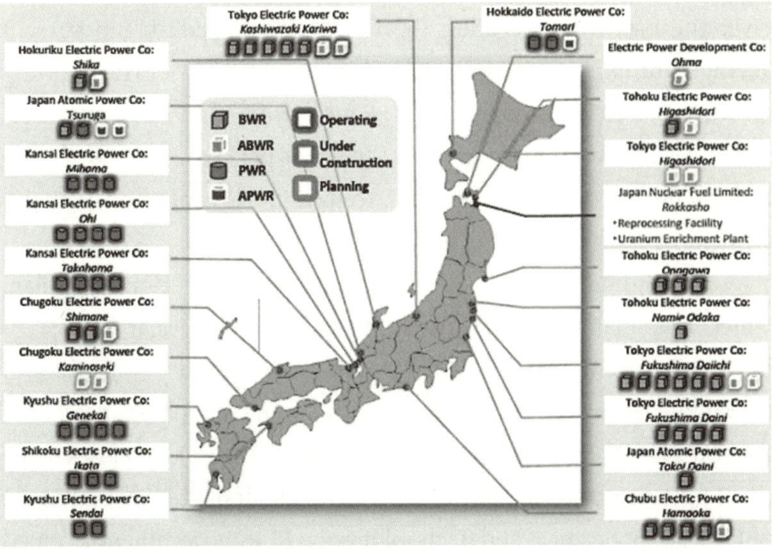

Nuclear Power in the Republic of Korea

The ROK Nuclear Policy

Fossil-fuel resources are very limited in South Korea. In 1957, the Republic of Korea (ROK) became a member of the International Atomic Energy Agency, initiating its nuclear activities. The following year, the ROK government passed the Atomic Energy Law and established the Office of Atomic Energy in 1959. In 1962, ROK's first research reactor achieved criticality. The first commercial-scale nuclear power plant, Kori-1, a Westinghouse unit built on a turnkey contract, started up in 1977 and achieved commercial operation in 1978. After that there was a burst of activity, with eight reactors under construction in the early 1980s.

The ROK's energy policy has been driven by considerations of energy security and the need to minimize dependence on imports. Over the past three decades, the ROK has enjoyed an 8.6 percent average annual growth in GDP, which has caused corresponding growth in electricity consumption. Power demand in the Republic of Korea has increased by more than 9 percent per year since 1990, and ROK's energy policy is to continue to have nuclear power as a major element of electricity production.

According to ROK's fifth long-term power-development plan, which was finalized in January 2000, eight more nuclear units were to be constructed by 2015, which would bring nuclear to one-third of the country's total generating capacity and would supply 45 percent of the country's electricity. Under the third comprehensive nuclear-energy development plan (for 2007–11), which was drafted by the Ministry of Education, Science, and Technology, the ROK government planned to develop its nuclear industry into one of the top five in the world, with about 60 percent of electricity from nuclear by 2035. In November 2011, the government reaffirmed its commitment to nuclear energy and targeted completion of six new reactors by 2016. The Ministry for Knowledge Economy announced plans for 59 percent of domestic electricity to be from nuclear by 2030 and said South Korea should be the third-largest reactor exporter by 2030, supplying 20 percent of the market, under a plan known as Nu-Tech 2030.[11]

Nuclear Power in the ROK

Compared with other countries, the ROK has a relatively small number of nuclear generating stations. There are only four nuclear-power generating stations, but each plant houses four or more units, and three sites have more reactors planned. So the ROK's nuclear-power

---

[11] *Xinhua Daily*, November 23, 2011.

production is more centralized than that of most nuclear-power nations. Though housing multiple units at each site allows more efficient maintenance and lower costs, it also reduces grid efficiencies.

The first three commercial units–Kori 1 and 2 and Wolsong-1–were bought as turnkey projects. The next six, Kori 3 and 4, Yonggwang 1 and 2, Ulchin 1 and 2, comprised the country's second generation of plants and involved local contractors and manufacturers. In the mid 1980s the Korean nuclear industry started to standardize the design of nuclear plants and to achieve much greater self-sufficiency in building them. In 1987, the industry selected the CE System 80 (2-loop) steam-supply system as the basis of standardization and then incorporated many of the US advanced light-water reactor design requirements. The result is the the Korean Standard Nuclear Plant (KSNP), which marked significant technical independence for Korea. In the late 1990s, to meet evolving requirements, the ROK started a program to produce an Improved KSNP, or KSNP+, and Shin-Kori 1 and 2 represented the first units of the KSNP+ program. This Generation II design is being offered for export as the Optimized Power Reactor–OPR-1000.

Beyond this, the Generation III Advanced Pressurised Reactor-1400 (APR-1400) draws on CE System 80+ innovations, which are evolutionary rather than radical. The basic design of APR-1400 was completed in 1999 and design certification was awarded in May 2003 by the Korean Institute of Nuclear Safety. It offers enhanced safety, with seismic design to withstand 300 Gal. ground acceleration, and has a 60-year design life. The cost is expected to be 10 to 20 percent less than KSNP/OPR-1000. Korean government data is reported to put the overnight cost of APR-1400 at the end of 2009 as $2300/kW, compared with $2900/kW for EPR and $3580/kW for the GE Hitachi ABWR. The same data puts the generation cost for Areva's APR at US$ 3.03 cents per kilowatt-hour, compared with an estimated 3.93

cents/kWh for EPR, and 6.86 cents/kWh for ABWR.[12] The first APR-1400 units—Shin Kori 3 and 4—were authorized in 2006 and were expected to be operational in 2013 and 2014.

Furthermore, in 2007, the ROK decided not to renew its reactor-technology licence agreement with Westinghouse but to embark upon a business cooperation agreement instead, which aimed to develop an exportable generation III+ APR+ large (1500 MWe net) reactor design by 2015.

For a long time, South Korea had been considered the main future market for nuclear-power expansion. In December 2009, South Korea stunned international observers when, in its first major overseas nuclear deal, the ROK succeeded in "snatching" a reported $20 billion contract with the United Arab Emirates (UAE) from the world's largest builder, AREVA, backed by the French state utility EDF ("Team France"), to build four 1.4 GW reactors. The first unit is scheduled to come on line on May 1, 2017.[13] Shortly following its sale of four modern nuclear-power reactors to the United Arab Emirates (UAE), the South Korean Ministry of Knowledge Economy declared in January 2010 that it aimed to achieve exports of 80 nuclear-power reactors worth $400 billion by 2030, in the course of becoming the world's third largest supplier of such technology, with a 20 percent share of the world market, behind the United States and France or Russia.[14] Following the UAE sale, the ROK is also pursuing opportunities in Turkey and Indonesia, as well as in Southeast Asia.[15]

---

[12] World Nuclear Association, "Nuclear Power in South Korea"; http://www.world-nuclear.org/info/default.aspx?id=348&terms=nuclear%20in%20korea.

[13] "UAE's First Nuclear Power Station to Start in 2017," *Energy Tribune*, December 22, 2010.

[14] "ROK Government Will Export 80 Nuclear Reactors in Next 20 Years," *Korean JoongAng Daily*, January 14, 2010.

[15] David Adam Stott , "South Korea's Global Nuclear Ambitions," *Asia-Pacific Journal*, March 22, 2010.

Figure 10. ROK NPPs Distribution

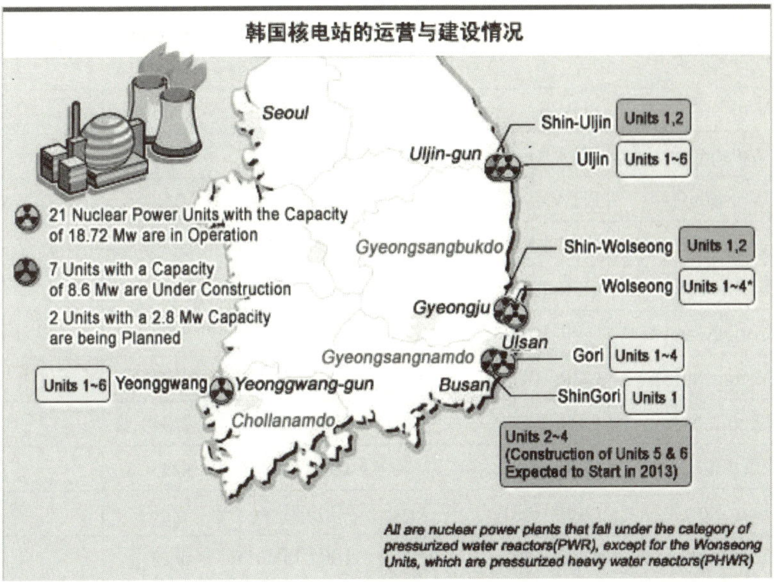

Table 7. Power Reactors Operating in South Korea

Reactor	Type	Net capacity operation	Commercial Close	Planned
Kori 1	PWR – Westinghouse	576 MWe	4/78	2017
Kori 2	PWR – Westinghouse	637 MWe	7/83	
Wolsong 1	PHWR – Candu 6	666 MWe	4/83	2036
Kori 3	PWR – Westinghouse	1007 MWe	9/85	
Kori 4	PWR – Westinghouse	1007 MWe	4/86	
Yonggwang 1	PWR – Westinghouse	953 MWe	8/86	
Yonggwang 2	PWR – Westinghouse	947 MWe	6/87	
Ulchin 1	PWR – Framatome	945 MWe	9/88	
Ulchin 2	PWR – Framatome	942 MWe	9/89	
Yonggwang 3	PWR (Syst 80)	997 MWe	12/95	
Yonggwang 4	PWR (Syst 80)	994 MWe	3/96	

Table 7. **Power Reactors Operating in South Korea (Continued)**

Reactor	Type	Net capacity operation	Commercial Close	Planned
Wolsong 2	PHWR – Candu	710 MWe	7/97	
Wolsong 3	PHWR – Candu	707 MWe	7/98	
Wolsong 4	PHWR – Candu	708 MWe	10/99	
Ulchin 3	OPR-1000	994 MWe	8/98	
Ulchin 4	OPR-1000	998 MWe	12/99	
Yonggwang 5	OPR-1000	988 MWe	5/02	
Yonggwang 6	OPR-1000	996 MWe	12/02	
Ulchin 5	OPR-1000	1001 MWe	7/04	
Ulchin 6	OPR-1000	1001 MWe	4/05	
Shin Kori 1	OPR-1000	1001 MWe	2/11	
Shin Kori 2	OPR-1000	1001 MWe	6/12	
Shin Wolsong 1	OPR-1000	1001 MWe	6/12	
Total: 21		20,787 MWe		

* Net capacities have been updated from PRIS January 2011 and Wolsong 1 in July 2011, but for the last five OPR units there is some confusion on net output. The KHNP Web site shows actual gross around 1055 MWe.[16]

Table 8. **South Korean Reactors under Construction, on Order, or Planned**

Reactor	Type	Gross capacity	Start construction	Commercial operation
Shin Wolsong 2	OPR-1000	1000 MWe	September 2008	1/2013
Shin Kori 3	APR-1400	1350 MWe	October 2008	9/2013
Shin Kori 4	APR-1400	1350 MWe	August 2009	9/2014
Shin Ulchin 1	APR-1400	1350 MWe	Sept 2012	4/2017
Shin Ulchin 2	APR-1400	1350 MWe	Sept 2013	4/2018
Shin Kori 5	APR-1400	1350 MWe	8/2014	12/2018

---

16 World Nuclear Association, "Nuclear Power in South Korea."

Table 8. **South Korean Reactors under Construction, on Order, or Planned (Continued)**

Reactor	Type	Gross capacity	Start construction	Commercial operation
Shin Kori 6	APR-1400	1350 MWe	8/2015	12/2019
Shin Wolsong 3	APR-1400	1350 MWe		6/2020
Shin Wolsong 4	APR-1400	1350 MWe		6/2021
Total: 9		**12,200 MWe** (c 11,680 MWe net)		

* Those not under construction are listed as planned in the WNA reactor table. The dates given in bold indicate plants under construction.[17]

The Future of Nuclear Power in Northeast Asia

Nuclear power has slowly been declining in recent years. According to the IAEA, some countries are now past their nuclear peak. Foe example, the three phase-out countries (Italy, Kazakhstan, and Lithuania) and Armenia generated their historical maximum of nuclear electricity in the 1980s. Several other countries (Belgium, Canada, Japan, and the United Kingdom) had their nuclear—power-generation peak in the 1990s. And seven more countries (Bulgaria, France, Germany, India, South Africa, Spain, and Sweden) peaked between 2001 and 2005. Between 2008 and 2009, nuclear generation declined in four of the "big six" countries—France, Germany, South Korea, and the United States. In Japan and Russia, the production remained just stable. These six countries generate nearly three-quarters (73 percent in 2009) of the world's nuclear electricity. In 2009, nuclear-power plants generated 2,558 TWh of electricity, about 2 percent less than the previous year.

---

[17] World Nuclear Association, "Nuclear Power in South Korea."

Just as the World Nuclear Association said: "another drop in nuclear generation"—the fourth year in a row.[18]

But by contrast, some other countries' nuclear generation—such as that for China, the Czech Republic, and Romania—showed a remarkably steady increase. With 18 of the 31 reactors now being built located in Asia and more in the planning, Asian nations are at the forefront of the renewed interest in nuclear-power generation, according to the findings of a released IAEA report, "Energy, Electricity, and Nuclear Power for the Period up to 2030" The report identifies countries such as Japan, South Korea, China, and India as the center of a current global nuclear-power expansion.[19] It is factors such as growing energy demand, energy security, and environmental concerns that are driving the Asian nuclear-power expansion. China and India have booming economies, booming populations, and growing energy demand, and they basically need to develop all the energy sources they can. In Japan and South Korea, the problem is not so much booming populations as a lack of indigenous oil and gas resources, and for them nuclear is attractive for energy-security reasons and—in particular in Japan—to reduce greenhouse-gas emissions. But is this a true picture of what is happening? And what is the future of nuclear power in northeast Asia?

Rethinking the Role of Nuclear Power after Fukushima

The March, 2011 disaster at the Fukushima Daiichi nuclear power plant in Japan has renewed concerns over the safety and viability of nuclear energy in the world. The concerns are timely as countries seek

---

[18] World Nuclear News, May 5, 2010.

[19] "Asia Leads Way in Nuclear-power Development"; http://www.iaea.org/newscenter/news/2007/asialeads.html.

to reduce carbon-intensive fossil-fuel based energy production while meeting growing energy demands through clean energy. With growing concern over rising fossil-fuel emissions and the lack of reliable renewable power, nuclear power is seen by many as a clean and stable source of power. But while nuclear power may be thought as a viable and necessary source of energy, the earthquake and the resulting nuclear disaster in Japan reminded the world about the inherent dangers of nuclear power—as did other past events, at Chernobyl and Three Mile Island, which were caused by mechanical failures and human error. The dangers of nuclear power include flaws in design and structural integrity, and age—factors that can threaten major metropolitan areas in the case of malfunction, economic feasibility, relicensing, storage, long-term disposal of highly radioactive waste, proliferation, and terrorist attacks.

If we add up the health, economic, environmental, and proliferation risks implicit in nuclear power, we can easily find that nuclear technology does not meet the criteria for a sustainable energy future. Below I list reasons for this.

Nuclear-Power Plants Are Not Good Economics

Nuclear-power plants are unable to compete on the open market without large government subsidies. Constructing a new nuclear-power plant can cost as much as $12 billion dollars and may take decades to produce a financial pay-off. According to the *Green Scissors Report*, since the 1940s the nuclear industry of the United States has received more than $125 billion dollars in federal subsidies for research and development, construction, clean-up, and insurance.[20] Even at the low-end, subsidies account for a significant portion of

---

[20] *Green Scissors Report*, 2010; http://www.greenscissors.com/GreenScissors2010.pdf.

nuclear power's operating cost advantage over competing energy sources.[21] And even with the huge subsidies, the annual renewables capacity additions have outpaced nuclear start-ups for 15 years. In the United States, the share of renewables in new capacity additions skyrocketed from 2 percent in 2004 to 55 percent in 2009. In 2010, for the first time, worldwide cumulated installed capacity of wind turbines (193 gigawatts), biomass and waste-to-energy plants (65 GW), and solar power (43 GW) reached 381 GW, outpacing the installed nuclear capacity of 375 GW prior to the Fukushima disaster.[22]

Problems in Relicensing Aging Reactors

Licenses are required to operate a nuclear-power plant in order to ensure adequate protection of public health and safety. In normal conditions, licenses are granted for a set period of time, usually 20 to 40 years, after which the nuclear operator must secure another license for continued operation. But as nuclear reactors near the end of their expected operating life, many owners apply for license extensions to operate reactors for an additional 20 years in order to avoid the expensive process of decommissioning a nuclear plant or building a new one.

The relicensing of old nuclear plants compromises the safety of these facilities and of the people who live close to them. Many of the problems that have been documented at existing plants, such as mechanical failure and issues of structural integrity, will be exacerbated if these plants are allowed to operate past their initial license date.

---

21 "Nuclear Power Subsidies: The Gift that Keeps on Taking"; http://www.ucsusa.org/nuclear_power/nuclear_power_and_global_warming/nuclear-power-subsidies-report.html.

22 World nuclear industry status report, 2010–2011.

Health Issues and Environmental Destruction

Nuclear-power plants generate power by splitting apart atoms, which produces radioactive waste. The radioactive waste is one of the most serious and fatal health issues involved in nuclear-power production. During a nuclear accident, the risk is multiplied because large amounts of radioactive material can be emitted into the atmosphere and carried hundreds of miles, depending on the wind and ocean currents. Exposure to large amounts of radiation can kill or mutate cells of living organisms. A comprehensive study of health effects of the 1986 Chernobyl accident by the World Health Organization found that incidences of thyroid cancer, leukemia, and cataracts increased greatly. Furthermore, radiation contamination can last for decades, causing prolonged health risks over vast amounts of once-populated areas. For example, the nuclear accident at Chernobyl contaminated 125,000 square miles in Belarus, Ukraine, and Russia, and spread radioactive material as far as the British Isles, 1,500 miles away, and caused significant environmental damage to native species and agricultural crops.

But people always misunderstand the environmental impacts of nuclear energy. Though the nuclear plants emit a limited amount of carbon dioxide in their end product—electricity—the industries that supply the raw materials for nuclear-power plants release a lot of carbon, which contributes to global climate change. The production of uranium, which powers most nuclear plants, is one of the most carbon-intensive industrial operations, once the mining, milling and chemical conversion processes are factored in. So nuclear energy is not so clean.

Storage and Disposal of Nuclear Waste

Waste from a nuclear reactor is highly radioactive and must be disposed of in a way that protects people and the environment from exposure and contamination. While the radioactivity of waste does diminish over time, it can continue to be dangerous for several hundred thousand years. Nowadays, there are two kinds of method, interim storage and long-term storage, to tackle the problem, but unfortunately none of them is a permanent solution.

Nuclear waste is usually stored and cooled in pools of water. This kind of storage pools is vulnerable to both accidents and intentional attacks that could lead to a loss of pool water, which will cause the used fuel rods to catch fire and melt, which releases radioactive material. The miserable disaster of the Fukushima nuclear plant is the best example. Given the vulnerabilities of storage pools, a better solution would be to move spent fuel into dry casks. What to do with the casks after they reach their expected life, however, is an unsolved problem. Initially, there is interim storage, but, while for the long term many options have been proposed, because spent fuel can stay radioactive for hundreds of thousands of years, the consensus is that nuclear waste should be stored in underground in "geologic" repositories. The risk of there being any repository depends on the geochemistry of the site, which can govern the chances of seismic activity, leakage, and groundwater contamination. There are also significant political hurdles to overcome when choosing a site. Without a viable option for disposal of this waste, the risks to citizens and the environment grows day by day.

Nuclear Proliferation

There is no way to completely separate the nuclear technologies employed for 'peaceful' purposes from those employed in weapons. Nuclear plants produce plutonium that can be used to form the core of a nuclear bomb. Plutonium and uranium can be separated from nuclear waste and reused at the plant, but that plutonium can also be dangerous: separated plutonium is much easier to steal and employ in a nuclear weapon. It takes less than 20 pounds of plutonium to build a small nuclear bomb. This increases the risk of nuclear proliferation and undermines attempts to reduce the risks of nuclear war.

After Fukushima: The Future of Nuclear Power

What happened in the Fukushima Daiichi nuclear power plant caused a fierce debate on nuclear energy and had wide ranging effects on countries that are pursuing and researching nuclear power. They began to rethink and reassess the developing method and some of them changed their nuclear policy. Others abandoned nuclear-power plants foreever.

Following the Fukushima nuclear disaster, Germany has permanently shut down eight of its reactors and pledged to close the rest by 2022.[23] The Italians have voted overwhelmingly to keep their country nonnuclear.[24] Switzerland and Spain have banned the construction of new reactors.[25] Taiwan's president did the same. Mexico has sidelined construction of ten reactors in favor of developing natural-gas-fired plants.[26] Belgium is considering phasing out its nuclear plants, perhaps as early as 2015. But as of April 2010, there were 112 nuclear power reactors operating in Asia, 37 under construction, and firm plans to build another 84. Besides China, which accounts for two-thirds of all reactors under construction, South Korea, India, Japan, Taiwan and Pakistan are also building new reactors. Power

---

23 Annika Breidthardt, "German Government Wants Nuclear Exit by 2022 at Latest," *Reuters*, May 30, 2011.

24 "Italy Nuclear Referendum Results," June 13, 2011; http://referendum.interno.it/referendum/refe110612/RFT0003.htm.

25 Henry Sokolski. "Nuclear Power Goes Rogue," *Newsweek*, November 28, 2011.

26 Carlos Manuel Rodriguez. "Mexico Scraps Plans to Build 10 Nuclear Power Plants in Favor of Using Gas," *Bloomberg Businessweek*, November 4, 2011.

reactors are at their planning stage in Bangladesh, Indonesia, North Korea, Thailand, and Vietnam. What will be the impact of the recent nuclear disaster in Japan on northeast Asia's nuclear future?

China—nuclear power's largest prospective market—on the one hand repeatedly assured the Chinese public that the radiation from the Fukushima nuclear accident would not have an impact on China and emphasized that China embraced nuclear power at a point when the technology had been developing in other countries for decades; therefore, China's nuclear-power plants were established with advanced levels of technology, so that China's nuclear facilities were all safe and better managed than those in Japan. On the other hand, the Chinese government held a State Council meeting to discuss the Japanese nuclear crisis and to consider China's own nuclear planning. At the meeting the government made three major decisions on nuclear power. First, the government decided to halt its plan to build new nuclear power plants. Second, it ordered a reexamination of the safety risks of nuclear-power stations currently under construction. Any safety faults discovered would lead to construction being stopped. Third, a decision was made to enhance safety measures in nuclear-power stations currently in operation in China.[27] In a rare stand, the Chinese government indicated that the utmost priority should be given to nuclear safety. Within three months ,the inspections of operating plants had been completed, and those on plants under construction were completed by October, some supplementary safety measures were announced, and a new China National Plan for Nuclear Safety, with short-, medium-, and long-term actions is being formulated. At the same time, a series of research and development (R&D) projects was launched by the NEA in February 2012 to improve safety-related technology and the country's emergency-response capabilities at

---

[27] *Xinhua Daily*, March 16, 2011.

indigenous nuclear-power plants in the event of an extreme disaster beyond the design bases.

Though China's State Council meeting decided to temporarily halt rampant nuclear growth, the prospects of slowing down the entire sector are not good. At best, their decisions would halt the development of some inland nuclear-power plants that are challenged by water shortages. Nuclear-power plants built inland generally use river water for cooling, and nuclear accidents can thus lead to the pollution of water supplies to regions with dense populations. But given China's thirst for energy to fuel economic growth, which is the basis for continued regime legitimacy, China is more than likely to continue moving ahead rapidly to build new nuclear-power plants. Particular emphasis will be given to Generation IV high-temperature, gas-cooled reactors with pebble-bed fuel, which have improved operating safety. In a word, though the Chinese government may hold off on nuclear-energy expansion, China is unlikely to completely abandon its current nuclear-power strategy.

Of the three northeast Asian countries, Japan was the first to pursue a nuclear-power strategy. Japan is energy poor and spends more than 86 percent of its imports on oil, gas, and other fossil fuels. This is what led to an increasing dependence on nuclear energy, which did supply almost 20 percent of the nation's energy needs before the earthquake and is one of the reasons why Japan built more power stations after the Chernobyl disaster.

But after the Fukushima disaster, public opinion polls showed that more and more people want to decrease the use of nuclear power, and Japan's prime minister has called for a dramatic reduction in Japan's reliance on nuclear power.[28] Given this trend, it is likely that

---

[28] Tsuyoshi Inajima and Yuji Okada. "Nuclear Promotion Dropped in Japan Energy Policy After Fukushima," *Bloomberg*, October 28, 2011.

the consumption of nuclear power in Japan will drop significantly over the next decade. Existing reactors, particularly those close to fault lines, will be off-line for extended periods and some will be decommissioned. It is expected that Japan's plans to build new reactors will be downsized, but the public must first be convinced that reactors are able to withstand earthquakes and tsunamis. To make up for lost nuclear power, in the short term, Japan is likely to import more coal and gas from Australia and Indonesia. In the longer term, Japan will significantly increase investment in alternative and renewable energy.

In South Korea, after the Fukushima accident there was immediate assessment of each site followed by a special safety review of all plants by the Ministry of Education, Science, and Technology, with special attention to Kori-1. This was followed by an IAEA Integrated Regulatory Review Service check of the whole ROK situation. A number of measures were initiated: the coastal barrier at Kori-1 was raised to 10m.; watertight doors were fitted to emergency diesel-generator buildings; battery-power supplies were secured from the possibility of flooding; a vehicle with a portable diesel generator was situated at each site; pumps were waterproofed; passive hydrogen removal systems not dependent on electricity were installed; exhaust and decompression equipment was improved; and the seismic performance of automatic shutdown and cooling systems was improved. All this represents an investment of about US$1 billion over five years. Given the extent of recent investment in the future of nuclear power, the ROK is likely to continue building new reactors, albeit at a slower pace. Emphasis will be given to upgrading safety features of the already operational reactors and to improving the safety features of the ROK's new APR1400 reactor.

* * * * *

After the Fukushima accident, the future of the expanding nuclear industry may have reached its crossroads. But there is little debate as to what energy can replace it, despite growing evidence that geothermal, hydroelectricity, and wind and solar energies are becoming more efficient, and less costly. In an increasingly energy-dependent world, the solutions to how to maintain the energy needs of seven billion plus people with a decreasing supply of nonrenewable energy supplies like oil still look unclear. The options available between using nuclear energy or renewable energy still pose divisive questions.

If we take into consideration all the health, economic, environmental, and proliferation risks implicit in nuclear power, the technology does not meet the criteria for a sustainable energy future. To meet present and future energy demands, we must invest in truly clean and renewable energy sources that can move the world toward a safe and healthy carbon-free economy.

Chapter 7

Overcoming the Cold War Mindset: A Case for Promoting a Nuclear-Weapons Convention and a Nuclear-Weapon-Free Zone

Takao Takahara

Don't be afraid to take a big step if one is indicated. You can't cross a chasm in two small jumps.

– David Lloyd George

Fukushima

The Fukushima nuclear crisis cast light on the existing threat posed by nuclear-power plants. Even expert critics who very well knew that "terrible things could happen" were struck by the severity of the accident. Many of them confessed that this was "beyond all projection." Knowledge had coexisted with complacency for a long time. The threat had continuously been with us for years when that major earthquake struck on March 11, 2011.

For people in Japan, where we realize that we live in a country where earthquakes are becoming even more frequent, the ongoing crisis has inevitably brought about a major review of all the existing fifty-some nuclear reactors that generate electricity. Japan is undergoing a societal self-education. In addition to the dangers from accidents, the very fact that the whole technology inevitably threatens the environment and health of workers in everyday operations is being more widely recognized. There also is a growing concern over the spread of nuclear-power plants to countries where people have so far been detached from nuclear technology. Reflecting the majority view that supports denuclearization, the government came out with a policy on September 14 that aims toward reaching that goal by the 2030s. There is die-hard opposition from the establishment, however, and the future seems somewhat uncertain.[1] The rift between the popular view and that of the establishment is serious. Constructive debate is still called for.[2]

At the same time, through the crisis, many Japanese were reminded of the perennial threat posed by nuclear weapons.[3] Indeed, perception of that threat had been more widely shared during the postwar confrontation between the two camps each led by nuclear superpowers. But after the "end" of the Cold War, one cannot deny that the public had drifted toward complacency. Now we should be able to realize better that the threat still remains with us at this very moment. Revisiting the question seems to be a must.

---

1 See, for example, an article by Stephen Hesse, "Japan's Nuclear Phaseout: Is It All Smoke and Mirrors?" *Japan Times*, September 23, 2012.

2 Although not complete, disclosure of information on the accident proceeds and should provide certain basis for the debate. In December 2011, the Diet (Parliament) of Japan organized an Independent Investigation Commission, which issued a rather critical comprehensive report on July 5, 2012. The report is available in English on the Web at http://naiic.go.jp/en/report.

3 As Tatsujiro Suzuki, acting chairman of JAEC, concluded his Dorothy Hodgkin lecture july 2011, "if we cannot control nuclear power plants, how could we ever control nuclear weapons?" July 4, at the 2011 Pugwash Conference on Science and World Affairs, Berlin.

Nuclear Peril

By sheer luck, the world barely survived the Cold War "balance of terror." It is a common knowledge of history that the nuclear powers did actually consider using nuclear weapons at certain junctures of postwar confrontations.[4] Had the restraint that worked in each case been removed, we might have observed mushroom clouds looming over many incinerated ground zeros. Worse, there was no guarantee that such use would have been confined to small-scale "limited" wars. Exchanges of these weapons, whose destructive power are much greater than those primitive bombs dropped in 1945, might well have devolved into a catastrophic all-out war between the two camps.

This perilous past is not yet behind us. Confrontation involving the possible use of nuclear weapons exists today. This is especially true in Northeast Asia. In addition to the situation between Beijing and Taipei that still entails military aspects that could lead to devastation, legally a state of war persists on the Korean peninsula. In addition, we observe signs for possible recourse to use or threat of use of force in international border disputes and skirmishes over fishing or mining rights in the sea.

We tend to forget, but actually at the global level, the state of "overkill" still exists: there are some 19,000 nuclear warheads. Signifi-

---

[4] It is known that the United States considered the use of nuclear weapons on the following occasions: against Soviet troops in Iran in 1946; during the Berlin blockade in 1948; in Urguay in 1948; in Guatemala in 1954; during the French defeat in the war in Indochina in 1954; in the crisis in Lebanon, 1958; in crises in the Taiwan Strait in 1954 and 1958; in the crisis in Berlin, 1961; in the Cuban missile crisis of 1962; to rescue marines in Ke Sang in the Vietnam War in 1968; against North Vietnam from 1969 to 1972; against the DPRK in 1968; against the Soviet Union from 1969 to 1973; against Syria in 1970; in the Iranian crisis of 1980; against Iraq, from the time of the George H. W. Bush administration until 2003: Peter Kuznick, "Perception of the Americans on Dropping of the Atomic Bomb," in Akira Kimura and Peter Kuznick, *Hiroshima Nagasaki e no Enbaku Touka Saikou (Revisiting the Atomic Bombings of Hiroshima and Nagasaki)* (Kyoto, 2011), pp. 108–109.

cant numbers of them are known to be in an operationally ready state and could be launched within minutes, constantly posing the danger of an accidental nuclear exchange that no one wants.[5]

It is no wonder that two consecutive UN secretary-generals have placed nuclear disarmament as their "top priority." Kofi Annan, gave a lecture to Princeton University students on November 28, 2006. After listing worrisome global issues such as poverty, environmental degradation, infectious disease, and continuous interstate and intrastate conflicts, he pointed out that it is the existence of nuclear weapons that poses "the greatest danger" to humanity. Annan cited three reasons why he believed so: The existential threat caused by the utter destructiveness of these weapons, the current crisis that the international control system—the Nonproliferation Treaty (NPT)—faces, and an emerging threat from nuclear terrorism.

Annan's lecture was delivered close to the end of his serving term, but the next and current secretary-general, Ban Ki-moon, stood up against the five permanent members of the Security Council, all of whom are armed with nuclear weapons, in his first year of office. In his speech on October 24, 2008, he bravely declared that nuclear disarmament was something that must be achieved. In May 2010, when he appeared at a large NGO meeting on the occasion of the NPT Review Conference, he emphasized that "from the first day in office, I have made nuclear disarmament a top priority."

As with our UN secretary-general, the citizens of the world ought to be acutely aware about the danger that nuclear weapons pose. The sense of urgency, which is often emphasized by the mayors of Hiroshima and Nagasaki in their peace declarations in August, is currently not shared by the policymakers or the general world public.

---

[5] It was disappointing that this state was not ameliorated by the Nuclear Posture Review under Obama in 2010.

It is not only because of their old age that the survivors (and witnesses) of the atomic bombings—namely, the *hibakushas*—express their frustration about the world still not moving forward toward nuclear abolition. What needs to be questioned is the overall complacency that has come to dominate the nuclear-weapons issue following the end of the Cold War.

The cities of Hiroshima and Nagasaki have continuously assumed the role of reminding the world on the need to tackle the question quickly. The governor of Hiroshima launched the idea of inviting the NPT Review Conference to his city so that other countries' representatives, by visiting one of the sites of destruction and resulting tragedy caused by a single, primitive, nuclear weapon, would recognize the urgent need for action. Were that be too difficult—to gather hundreds of diplomats in the Far East—an alternative idea might be for the Japanese government to host Six-Party Talks in Hiroshima or Nagasaki. The Japanese government feels diffident about such a proposal, believing that it would displease the United States, but if that were overcome and if the Chinese government was willing, such an idea might well bring new impetus to the negotiations on denuclearization.[6]

An important factor is the huge cost of these weapons. According to the Stockholm International Peace Research Institute, there was a steep 5 percent annual increase in world military expenditure in the first decade of the 21st century. Although the rate of growth slowed in 2010, the annual expenditure in 2011 surpassed US$1.7 trillion.[7] The existence of nuclear weapons reinforces the premise behind arms

---

[6] The idea was also proposed by Professor Chung-in Moon in an op-ed article to the *Chugoku Shimbun*, January 25, 2010. An obstacle would be the Japanese government, which has a record of evading anything that might make the mighty ally uncomfortable. The humanitarian consequences of the atomic attack inevitably makes US citizens feel guilty, although that is definitely not survivors' intent.

[7] Stockholm International Peace Research Institute, SIPRI Yearbook, 2012: Armaments, *Disarmament and International Security* (Oxford University Press, 2012), p. 149.

build-up, that wars are possible and nations must be prepared. The budget allocated for nuclear weapons themselves is not exactly clear, but according to an analysis by a French journalist, the estimated amount in 2008, in US dollars, came somewhere close to between 89 and 93 billion.[8] Although this is less than one-tenth of total military expenditure, the amount spent daily is almost equivalent to the whole annual United Nations Environment Program (UNEP) budget.

More recently, NGOs such as International Physicians for the Prevention of Nuclear War (IPPNW) claimed that unexpectedly severe consequences of climate change are likely to occur after even limited use of nuclear weapons. Similar to the studies on "nuclear winter" in the early 1980s that warned about the environmental consequences that would ensure from a nuclear exchange between the two sides of the Cold War, IPPNW's 2012 research study explains how a nuclear war anywhere in the world, using as few as 100 weapons, would disrupt global climate and agricultural production so severely that the lives of more than a billion people would be at risk.[9] Such warnings from scientists have come together with renewed concern about the humanitarian consequences of any use of nuclear weapons—a topic to be mentioned in a later section.

The Obama Question

One tends to forget that the US president, Barack Obama, was the 2009 Nobel peace laureate. A major reason for his winning the prize was the initiative toward a "world without nuclear weapons." This

---

[8] Ben Cramer, *Nuclear Weapons: At What Cost?* (Geneva: International Peace Bureau, 2009).

[9] Ira Helfand, *Nuclear Famine: A Billion People at Risk—Global Impacts of Limited Nuclear War on Agriculture, Food Supplies, and Human Nutrition* (IPPNW, April 2012).

goal not only reflects his idealism but also is based on a realistic assessment of the situation we face—that is, stark recognition that nuclear peril today is serious enough to conclude that a global system supported by an institutional norm of "no nuclear weapons" would better serve the security of nations, including those currently armed with nuclear weapons. If the threat derives from clandestine storage of nukes, it would be easier to search for them where "zero" is the normalcy.

The United States' utter superiority in the conventional arsenal has been a favorable background for the recent US proposals for a nuclear zero. Zero was the conclusion of the four former Cold War warriors who together authored the celebrated op-ed piece in the *Wall Street Journal* in January 2007.[10] Received with surprise at the time of its publication, it was nevertheless taken seriously by an increasing number of US presidential candidates in 2008, and spread to discussions in the mainstream media, as well as being on the platform of the winner, Obama. This was a historic moment: Mainstream United States at last seemed to have taken up the issue of nuclear disarmament seriously. On April 5, 2009, Obama made a speech in Prague, that encouraged peace movements across the world. "As the only nuclear power to have used a nuclear weapon, the United States has a moral responsibility to act.... I state clearly and with conviction America's commitment to seek the peace and security of a world without nuclear weapons," he declared.

The questionable compromises that Obama later made with the more conservative politicians on Capitol Hill, such as allocating abundant investment to so-called nuclear infrastructure, might well have long-term adverse consequences for nuclear disarmament. In

---

[10] George P. Shultz, William J. Perry, Henry A. Kissinger, and Sam Nunn, "A World Free of Nuclear Weapons," *Wall Street Journal*, January 4, 2007. See also the articles by the same authors in later years.

the end, he could not even secure the support of a right-wing senator that he had sought. He did, however, conclude a new arms-control treaty with Russia, and won ratification. At the 2010 NPT Review Conference, the United States showed internationalist approaches and contributed immensely to evasion of the much-feared collapse of the nonproliferation regime through failure to agree on any documents at the conference. Both achievements in 2010 would have been difficult under the previous G. W. Bush administration. There are limits, however, to the Obama administration's approach. After the quote above in his speech in Prague, he quickly stepped back, saying, "Make no mistake: As long as these weapons exist, the United States will maintain a safe, secure and effective arsenal to deter any adversary, and guarantee that defence to our allies."

Obama's words show that his posture is different from traditional approaches to nuclear disarmament. Nuclear disarmament has been promoted from basically two perspectives. Firstly, was the emphasis that man must avoid bringing an end to the human race. This approach is exemplified, for example, in the famous (but often misunderstood) Russell-Einstein Manifesto of 1955, which appealed for a reconsideration of the notion of sovereignty in order to abolish war. According to this manifesto, nuclear disarmament must be promoted, but that will not be enough since the world now knew how to make these weapons. It calls for a totally new way of thinking, since the world has radically changed as it entered the nuclear age. "Remember your humanity and forget the rest" is the core message of the manifesto.

Secondly, there is an approach that emphasizes the inhumane and indiscriminate (hence illegal) character of the nuclear weapon. In a sense this approach is in the tradition of international law to prohibit certain types of weapons. From this perspective, no more Hiroshimas can be tolerated, even if the world may not be brought to an end by such use. This is closer to the feeling of the *hibakushas*, although they speak of "an end of the world" if the bomb is again used anywhere on Earth. Both these approaches deny all use of nuclear weapons, and usually they are not consciously distinguished. It is useful, however, to distinguish the two when discussing the policy implications of current antinuclearism.

Obama's approach does seek a "world without nuclear weapons," but the road to that goal is to be taken step by step, steadily reducing the role that nuclear weapons play in the realist world of international relations. As the quote from the Prague speech clearly shows, Obama sees that the United States must maintain its own nuclear weapons as long as these weapons exist. For the sake of its own security, the United States will be the last nation to abandon these weapons, he seems to say.

The logical outcome of this approach is obvious. As long as the United States retains its nuclear arsenal ready for use, the states that fear being possibly attacked by the United States will inevitably seek to maintain some deterrent against such attack. It is natural for them to desire possession of the same kind of weapons, if that is achievable, especially after witnessing the fate of Sadam Hussein and Muammar Ghadaffi. Looking back at the history of the nuclear age, following Hiroshima and Nagasaki, it is the fear of being attacked by nuclear weapons that has perhaps been the strongest motivation for striving to acquire such weapons. In fact, that precisely was the case with the scientists who arduously participated in the Manhattan Project early in the 1940s. Whether or not they might they have been more strongly motivated by genuine enthusiasm for the science, "completing it before the Nazis" must have worked as the best justification for devotion to developing this diabolic weapon. Similar reasoning or justification was to be seen in other nations that followed the path of going nuclear, and the logic remains the same for those nations that today have to envisage armed conflict with the United States or other nuclear-weapon states.

In order to overcome this "nuclear standard," akin to the "gold standard" on which finance experts once placed such essential importance, two myths must be questioned. The first myth has to do with the perceived military effectiveness of the atomic bombings. Since Hiroshima and Nagasaki coincided with the very last phase of the Asia–Pacific War, it is a popular belief that the atomic bomb was essential in bringing the war to an end, saving the lives of many US soldiers and liberating the Asian nations from Japanese militarism. But this notion leaves ample room for questioning. The majority of the Japanese historians who have closely studied the very last days of Imperial Japan in the Asia–Pacific War believe that the decisive factor at that time for the Japanese leadership's decision to surrender was not the dropping of the atomic bombs but the abrupt Soviet entry into the war, which came as a surprise to them.[11]

Secondly, the value of nuclear deterrence must be demystified. The notion of deterrence is used to explain the utility of nuclear weapons, and the very word itself—*deterrence*—has the connotation that these weapons are for protection and defence. This has worked well as a way to justify the nuclear arsenal to taxpayers. To the contrary, nuclear deterrence is essentially a threat to retaliate against the enemy in a case of attack, and the threat must be understood by the adversary to be real or "credible." A continuous readiness and arms build-up follow from this, which makes it very difficult to bring about the stable balance that it claims to achieve. Nuclear deterrence militarily is not defensive, but inevitably an offensive posture towards

---

[11] In English, see, for example, Tsuyoshi Hasegawa, *Racing the Enemy: Stalin, Truman, and the Surrender of Japan* (Cambridge, MA: 2005); Ward Wilson, "The Winning Weapon? Rethinking Nuclear Weapons in Light of Hiroshima," *International Security* 31, no. 4 (2007): 162–179.

a potential adversary.

In his landmark speech in October 2008 cited above, the UN's Ban Ki-moon pointed out that the doctrine of deterrence was "contagious." Indeed, the notion of "extended deterrence" has insinuated a false sense of security into the minds of the policymakers in nonnuclear weapon states, such as Japan.

The Japan Problem

At present, the divide between Japan's policy elites and civil society is not confined to the civilian use of nuclear power but exists also in the issue of nuclear weapons. The policy elites cling onto protection by the "nuclear umbrella," while the civil society and the majority of the public reject such "use" of nuclear weapons.[12] Japan's declared policy on nuclear weapons at this point is, 1, to promote nuclear disarmament in multilateral for a, and 2, to rely on the "extended deterrence" of the United States nuclear arsenal. Behind these two pillars the "three nonnuclear principles" of never possessing, producing, or introducing nuclear weapons on Japanese territory are held up as the national policy. In order to promote nuclear disarmament, it is the second pillar—namely, Japan's reliance on the so-called nuclear umbrella—that needs to be questioned, diminished, and eventually

---

12 In a poll taken on the 60th year since Hiroshima and Nagasaki, a majority of the Japanese thought that a country's international influence would not be enhanced by acquiring nuclear weapons (33 percent responded yes, 58 percent said no); 35 percent "believed in nuclear deterrence," while 50 percent did not. On the former question, South Koreans' response was 63 percent yes, 27 percent no; in China, 70 percent vs. 2 percent; in the United States, 57 percent vs. 40 percent; in Germany, 50 percent vs. 48 percent. On the latter question, South Koreans responded with 62 percent saying yes, 36 percent saying no; in China, 39 percent vs. 58 percent; in the United States, 39 percent vs. 58 percent; in Germany, 23 percent vs. 75 percent: *Asahi Shimbun*, August 6, 2005.

abandoned.

How is Japan's reliance on extended deterrence expressed? In December 2010, Japan revised its basic policy document on defence, which includes its posture on nuclear weapons. Revision of the former National Defence Program Guidelines of 2004 was due back in 2009, but because of the defeat of the long-ruling conservative Liberal Democratic Party (LDP) and a sweeping victory by the center-left Democratic Party of Japan (DPJ) in August that year, the release of the document was postponed. The National Defence Program Guideline for FY2011 and beyond (hereafter, the NDPG) was adopted by the cabinet on December 17, 2010. It stuck to the long-upheld "exclusively defence-oriented policy … not becoming a military power that poses a threat to other countries," which barely compromises the contradiction between the existence of the Self-Defence Forces with the sixth largest military budget in the world, on the one hand, and the pacifist 1947 constitution, on the other.

Reflecting the permanent budgetary crisis, reduction of Ground Self-Defence Forces, such as cutting the number of tanks from 600 to 400, was inevitable. The NDPG admits that "a full-scale invasion against Japan that will threaten its existence, such as a large-scale landing invasion, is unlikely to occur," but finds that there are new and complex "security challenges and destabilizing factors," including the rapid increase of military expenditures of all the countries surrounding Japan.

Simply put, the characteristic of the new policy document is twofold: a shift of the frontline from the north (Russia) to the south (China), and continuation of cooperative policies with the United States on missile defence. These features disappointed many supporters of the DPJ who had wished the new government might redirect the region toward demilitarization by probing a new cooperative relationship with China (and Russia, too, for that matter) and start to withdraw from the expensive Missile Defense (MD) system that the Chinese, especially, had been sensitive about. However, the new government in Japan which is dominated by conservative Liberal Democratic Party (LDP) would prefer otherwise.

With regard to nuclear weapons, the usual short statement on how Japan should counter nuclear threats was revised totally. Following a sentence on Japan's commitment to nuclear disarmament, the relevant part read: "As long as nuclear weapons exist, the extended deterrence provided by the United States, with nuclear deterrent as a vital element, will be indispensable. In order to maintain and improve the credibility of the extended deterrence, Japan will closely cooperate with the United States, and will also appropriately implement its own efforts, including ballistic missile defence and civil protection." The wording apparently corresponds to the content of the new 2010 Nuclear Posture Review (NPR) of the United States. Interestingly, while on the US side 2010 NPR could be read in the context of efforts to reduce the role of nuclear weapons, in the Japanese context the new wording rather attributes a larger "vital" role in the extended deterrence to nuclear weapons. Before this, the 2004 NDPG simply stated: "To protect its territory and people against the threat of nuclear weapons, Japan will continue to rely on the US nuclear deterrent." The world of nuclear weapons had been somewhat detached from the world of conventional defence: Japan had nothing to do with the former, and absolutely had no say. We are yet to see what these changes in the wording might imply.

In any case, the problem with the new 2010 NDPG and the thinking of the security elites behind it is that it does not proactively comply with the promises repeatedly made in the multilateral negotiations; namely, to reduce the role of nuclear weapons in pursuing national security. To the contrary, NDPG continues to declare that Japan relies on the extended deterrence provided by the United States and, as explained above, may even be observed to have assumed a stronger role for nuclear weapons in the defence of Japan. More concretely put, the wording in the revised NDPG seems to premise a larger role in nuclear weapons (of the United States) by implying that extended deterrence covers planned attack by other types of weapons, even conventional weapons.[13]

In July 2010, while the new NDPG was being drafted, the opposition LDP committed itself to reducing the "three nonnuclear principles," which has long been the basic national policy of Japan, to "two and a half principles." This means that US vessels and planes armed with nuclear weapons would be allowed to make transit visits to Japanese ports and airfields. The underlying thinking of this change dates back to the Cold War: Japan needs to let the US nuclear weapons physically enter Japanese territory in order for the military ties to be effectively useful. This belief can be traced back to the 1960s, when the Japanese prime minister at that time asked for a "nuclear umbrella" from the United States in order to protect Japan in case of any invasion from nuclear China. Unsuccessful effort by Japanese diplomats to hinder the retirement of nuclear-capable US cruise missiles in the same year, 2010, is another case in point. In this context, Japan actually provides an excellent excuse for the political forces within the United States against Obama's vision of the world without nuclear weapons.[14]

---

[13] This coincides with the notoriously obstructive attitude of Japanese participants in the International Commission on Nuclear Non-Proliferation and Disarmament (the Evans- Kawaguchi Commission) on the issue of limiting the role of nuclear weapons to the "sole purpose" of deterring only nuclear attacks from the adversary.

[14] See, for example, "Why We Don't Want a Nuclear-Free World: The Weekend Interview with James Schlesinger," *Wall Street Journal*, July 13, 2009.

To this day, the Japanese defence establishment seems to have been caught in the myth of the "nuclear standard" of international politics. The victim, in this myth, sees international politics almost exclusively in terms of military power and nuclear weapons; one will not look for alternative means to effectively "deter" the adversary from attacking. According to this thinking, only by either acquiring nuclear weapons or by relying on the extended nuclear deterrent of an ally, could anyone fulfil that function. Many in the establishment even seem to feel that it is a sort of a privilege for a nation to enjoy that kind of promise of protection from a mighty nuclear ally, while the majority of nonnuclear states from the beginning just do not pursue that path.

As the only nation to have suffered nuclear bombings, Japan claims that it will work toward nuclear abolition. Its leaders repeatedly spoke at the podium of the United Nations and elsewhere, introduced General Assembly resolutions calling for nuclear disarmament every year since 1994. Prime ministers make promises every year on the occasion of Hiroshima and Nagasaki ceremonies in August that they will strive for the elimination of these weapons. Japan's disarmament diplomacy, however, loses much of its integrity when it comes to discussing concrete steps to reduce the role of nuclear weapons. At this time, the credibility of the Japanese representatives on nuclear disarmament rest largely on the dignity of the *hibakushas* and the persistently strong public opinion in Japan for nuclear abolition.[15]

A huge discrepancy existed and still exists between the policy elites and the majority of Japanese public opinion. Inconsistency had

---

15 An opinion poll conducted by the *Mainichi Shimbun* in November 2007 is an example. Respondents were asked: "After the North Korean nuclear test, there have been talks by the high officials and LDP politicians that the Japanese, too, should discuss whether or not to go nuclear. Are you for Japan to acquire nuclear weapons?" Replying, 78 percent said no; 14 percent said yes. For male respondents, the figures were 74 percent to 18 percent; for females 82 percent to 10 percent.

occasionally been hidden by secrecy and blatant lying on the part of the government, especially on the issue of allowing transit of US nuclear weapons through Japanese ports and airstrips. If Japan is a democracy, this situation is unsustainable, unless we live in a world where the government policy is seen as different from "the Japanese posture" on nuclear issues.

A Nuclear-Weapons Convention and Proposals for a Nuclear-Weapon-Free Zone

If Japanese statesmen and stateswomen wish to be taken seriously about their commitment toward nuclear disarmament, the touchstone issues will be proposals for a Nuclear Weapons Convention and a Nuclear-Weapon-Free Zone, in addition to the moot nuclear-reprocessing program. The 2010 NPT Review Conference came out with a final document that expressed "deep concern at the catastrophic humanitarian consequences of any use of nuclear weapons" and referred to a "nuclear weapons convention." It also clearly stated that the conference resolved "to achieve the peace and security of a world without nuclear weapons." The question now is whether or not the governments, including the Japanese, are ready to accept and proceed with these commitments.

Actually, the Japanese government has since adopted the wording "the catastrophic humanitarian consequences of any use of nuclear weapons" in the UNGA resolutions that Japan itself sponsors. The next step is to start on negotiations toward a convention. Japan did question the legality of atomic bombings back in August 1945. The wartime letter that was delivered to the United States via Switzerland has been left unanswered. After the surrender and ensuing occupation by the Allies, the Japanese government never questioned the legality even during the deliberations at the International Court of Justice that came out with an advisory opinion in July 1996.[16]

As regards the illegality of any use of nuclear weapons, progressive initiatives are currently taken in the Red Cross community. The unusual press conference held by President Jacob Kellenberger in April 2010 seems to have had significant effect in the postures of countries like Switzerland, Austria, and Norway at the 2010 NPT Review Conference and ensuing UN meetings. In late November in 2011, the International Red Cross and the Red Crescent Movement adopted a strong resolution to proceed toward elimination of nuclear weapons. Along with the Norwegian and the Australian Red Cross, the Japanese Red Cross took part in leading the council to agree on such a resolution.[17]

Following such initiatives, in May 2012, at the first preparatory committee for the 2015 Review Conference held in Vienna, sixteen nations led by Switzerland came out with a succinct statement that summarized current approaches toward nuclear abolition.[18] The sixteen nations included all three groups of nations that have been proposing a ban on nuclear weapons; namely, the Non-Aligned Movement, the New Agenda Coalition, and the new Red Cross—inspired countries that place emphasis on the "catastrophic humanitarian consequences" that are expected from any use of nuclear weapons.

---

16 The Japanese government notoriously tried at first to dissuade and eventually to distance itself from the testimonies of the mayors of Hiroshima and Nagasaki in the Hague. Both mayoral statements clearly pointed out the illegality of any use of nuclear weapons.

17 Council of Delegates, 2011: Resolution 1 "Working Towards the Elimination of Nuclear Weapons," November 26, 2011; http://www.icrc.org/eng/resources/documents/resolution/council-delegates-resolution-1-2011.htm.

18 "Joint Statement on the Humanitarian Dimension of Nuclear Disarmament by Austria, Chile, Costa Rica, Denmark, Holy See, Egypt, Indonesia, Ireland, Malaysia, Mexico, New Zealand, Nigeria, Norway, Philippines, South Africa, Switzerland"; http://www.reachingcriticalwill.org/images/documents/Disarmament-fora/npt/prepcom12/statements/2May_IHL.pdf.

Although there are observations that the Japanese government may be approaching a point when it begins its support for at least the start of negotiations toward a NWC, it still refrains from joining the nations that currently lead the discussion on humanitarian aspects. In fact, Japan was not consulted at all in the initiative led by Switzerland in Vienna in May 2012, allegedly for the reason that inclusion of Japan would very likely have led to a weaker statement. In October, at the first committee meeting of the UN General Assembly, Japan was invited this time to join the new statement by the same group. Japan declined, along with Australia and some NATO countries. This was very disappointing to the Japanese NGOs and the *hibakushas*, especially after news reports had depicted where Japan stands today. To Yoshihisa Taue, the mayor of Nagasaki, who appealed for Japan to take back the denial and join the statement, the candid reply from the Japan's Foreign Ministry official was a clear "no" because of Japan's reliance on the "nuclear umbrella."

Since disarmament cannot be anything but a process, history shows that it is worthwhile to start negotiations. Negotiations will set a goal, which will be legitimized and shared over time if talks are conducted seriously. Establishing a norm that nuclear weapons must never be used again will inevitably entail discussion also on new security arrangements in the region, which could lead the way to denuclearize nations' security policies.

In this context, it is interesting to note that while the Republic of Korea (ROK) and Japan have been abstaining from the so-called Malaysian Resolution in UNGA, which calls for a follow-up to the 1996 International Court of Justice (ICJ) ruling and for starting the negotiations toward a NWC, surprisingly North Korea has been voting positively on this.

As for a Nuclear-Weapon-Free Zone in Northeast Asia (hereafter, NEANWFZ), again to the disappointment of the NGO community the Japanese government so far has been stubborn to state that "it is too early to discuss it." It is well known that the United States so far does not like the idea. The prospective first step toward denuclearizing the current security arrangement with the United States seems to be accepting the idea of "sole purpose." It matches with the current Japanese policy of adhering to the "three nonnuclear principles" and also with the 1991 joint declaration of the ROK and North Korea, which the six party talks aims to bring back.

In December 2009, the then foreign minister of Japan, Katsuya Okada, sent a letter to the secretaries of state and defense of the United States. Among other things, he referred to his strong interest in the negative security assurances (NSA) that can be provided for the non-nuclear states. NSA is a component of a nuclear-weapon-free zone that could substitute the extended deterrence of a nuclear ally. Okada stepped down as foreign minister when Prime Minister Hatoyama resigned over the issue of the Futenma Marine Air Base in Okinawa. Okada's move did not bear fruit immediately, but it did show that policy initiatives are possible even within the existing security arrangement with the United States.

If North Korea now speaks of its own nuclear deterrent, doing away with the nuclear deterrent on the part of the ROK, and also Japan, might be necessary in order to bring about denuclearization of the Korean peninsula. Foreseeing a NEANWFZ could have positive meaning since it could pave the way for the three parties—namely, the DPRK, ROK, and Japan—to share understanding and a sense of a shared agenda for common security in the region. Northeast Asia is a region of concentration of nuclear-weapon states—namely, the United States, China, and Russia. If all three could agree on NSA with the three nonnuclear states, we will see a more stable situation. Possibly a new multilateral security framework will emerge in the region.

There is a window of opportunity at this moment in history.

There is a US administration committed to a world without nuclear weapons, and also with an undisputable superiority in conventional armaments. If we close this opportunity, it would be a huge loss not only for the region but also for the world, for there is urgent need for renewed impetus toward a world without nuclear weapons. Nuclear weapons need to be put off alert as soon as possible. The nuclear-weapon states should fulfil their promise to proceed toward total elimination. All chances of an unwanted nuclear exchange must be minimized and eliminated, for any use of nuclear weapons will mean a nuclear hell. And nations must be reminded that some nonnuclear states that have been complacent under the self-deceptive "nuclear umbrella" also are responsible for prolonging the precarious situation.

We should be aware that humanity is caught in a system that is outdated. As Albert Einstein presciently pointed out, everything has changed in the nuclear age except the way people think. The Russell-Einstein Manifesto of 1955 declared that "we have to learn to think in a new way.... Shall we put an end to the human race; or shall mankind renounce war?" The manifesto's prescription is all the more relevant today. Being aware of the perils of this ultimate weapon, the hope lies in our outgrowing the old war-system that we inherited from the preceding century.

Arguably, it is precisely because of the nature of the nuclear question that the elites who have vested interests in the current system are reluctant to proceed toward total abolition of nuclear weapons. If we straightforwardly tackle this nuclear question, we might be able to proceed to a new regional and global system. Starting negotiations toward NWC and NEANWFZ can be a beginning, and to proceed further cooperation between the two nonnuclear states of the ROK and Japan is a must.

The Contributors

Leon V. Sigal

Leon V. Sigal is director of the Northeast Asia Cooperative security Project at the social Science Research Council in New York. He was a member of the editorial board of the New York Times from 1989 to 1995.

Alexander G. Savelyev

Alexander G. Savelyev is Head of Strategic Studies Department, Center of International Security, Institute of world Economy and International Relations (IMEMO) at the Russian Academy of Sciences since 2002.

Sang-Hyun Lee

Sang-Hyun Lee is Director-General for Policy Planning, Ministry of Foreign Affairs and Trade (MOFAT). He was a research fellow at the Korean Institute for International Studies (1987–88), The Korea Institute for Defense Analysis (1988–90), and senior research fellow at the Sejong Institute (2001–2011)

Kun Young Park

Kun Young Park is a professor at the Catholic University of Korea, where he has taught International Relations since 1997. He is the author and co-author of over 50 scholarly articles, and author or editor of 5 books, including the International Politics of the Korean Peninsula and Contemporary Theories of International Relations and Korea.

Siegfried S. Hecker

Siegfried S. Hecker is co-director of the Stanford University Center for International Security and Cooperation, Senior Fellow of the Freeman Spogli Institute for International Studies, and Professor (Research) in the Department of management Science and Engineering.

Jian Cai

Jian Cai is an associate professor of international relations at Fudan University and serving as the executive director of Center for Korean Studies, director of the association of Korean history and director of the association of Sino-foreign relations, the executive edit-in-chief of the collected papers on Korean Studies.

Takao Takahara

Takao Takahara is Professor of International Politics and peace Research at Faculty of International Studies, meiji Gakuin University. He also is a member of the International Peace Research Institute Meiji Gakuin University (PRIME), to which he formerly served as Director (2002–2004).

The Editor

Su Hoon Lee

Professor Su Hoon Lee received his Ph.D. in comparative International Development and Sociology from Johns Hopkins University in 1986. In 1986, Dr. Lee joined Kyungnam University in Korea and simultaneously assumed the position of the Associate Director at its Institute for Far East Asian Studies (IFES) in Seoul. Since Feb. 2009, he serves as the Director of IFES, Kyungnam University. He has also served on various advisory committees of the ROK government; he was the Chair (Ministerial Level position) of the 'Presidential Committee on Northeast Asian Initiative' (2005–2007), and accompanied President Roh Moo Hyun for the 2007 inter-Korea summit held in Pyongyang as one of the special delegation. He is the author of, among others, *State-Building in the Contemporary Third World* (Westview Press, 1989), *World-System Analysis* (in Korea, 1993), *For a Humane World-System* (in korean, 1999), *Crisis and Capitalism of East Asia* (in Korean, 2001), and *World-System. Northeast Asia. and Korean Peninsula* (in Korean, 2004).

Index

[O]

[P]

[Q]

[R]

[S]

[T]

Explorations in Korean Studies

Books in the Series

THE DYNAMICS OF CHANGE IN NORTH KOREA
An Institutionalist Perspective (2009)

Edited by Phillip H. Park

LAW AND DEMOCRACY IN SOUTH KOREA
Democratic Development Since 1987 (2010)

Written by Dae-Kyu Yoon

NORTH KOREA
Development, Human Rights, and Democracy (2010)

Written by the North Korea Modernization Research Group

Nuclear North Korea
Regional Dynamics, Failed Policies, and
Ideas for Ending a Global Stalemate (2012)

Edited by Su Hoon Lee